AF757040

SEBASTIÃO SALGADO

GOLD

SEBASTIÃO SALGADO

GOLD

Serra Pelada Gold Mine · Goldmine Serra Pelada · *Mine d'or Serra Pelada*

Conception and design · Konzeption und Gestaltung · *Conception et réalisation*
LÉLIA WANICK SALGADO

TASCHEN

GOLD, AN AMAZON FEVER

When I first reached Serra Pelada, I was left speechless. Before me, I saw a vast hole, perhaps 200 meters in diameter and almost as deep, teeming with tens of thousands of barely clothed men, roughly half of them carrying heavy sacks up broad wooden ladders, the others leaping down muddy slopes back into the cavernous maw.

I of course knew what they were doing. They were looking for gold. And I had traveled to this remote hilltop on the edge of the Amazon rainforest in Brazil's northern state of Pará to photograph them. Even so, I was ill-prepared to take in this extraordinary spectacle.

What is it about a dull yellow metal that drives men to abandon their homes, sell their belongings, and cross a continent in order to risk life, limb and sanity for a dream? Serra Pelada—Bald Mountain, in Portuguese—was hardly the first such fever to strike the Americas. Almost five centuries earlier, the search for El Dorado drove the conquest of an entire continent. Landing on the Mexican coast in 1519, Hernán Cortés informed the Aztec emperor that his men suffered from an illness that only gold could cure. And the get-rich-quick stampede continued, reaching heights of madness in the California Gold Rush of the 1840s and Canada's Klondike scramble 50 years later.

No less than earlier waves of wide-eyed speculators, the prospectors, or *garimpeiros*, in Serra Pelada were infected by this same all-consuming zeal.

Gold was discovered in a stream there in 1979 and within weeks a gold rush had begun. The first men to arrive formed a cooperative, which assigned the pioneers two-meter by three-meter plots of land to prospect. Tens of thousands soon followed, becoming the peons who did the digging and carried the soil for the plot owners and their co-investors.

To maintain order, Brazil's military government sent federal police to the site, allowing work to continue but banning guns, alcohol, and women. Officially, the right to exploit this land belonged to the state-owned mining giant, Companhia Vale do Rio Doce (CVRD), now renamed Vale, but it was helpless to stop the frenzied freelance mining that was taking place on an almost-biblical scale.

I had tried to visit Serra Pelada in 1980, but was blocked by the restrictions on press and other freedoms imposed by Brazil's military when it seized power in 1964. It also didn't help that

I had chosen exile in Europe to protest the dictatorship. So for the next five years, I was refused access to the mine. Finally, in September 1986, with the military no longer in power and Pará's state police now in charge of security, the prospectors' cooperative gave me the green light to travel there.

By good fortune, a friend of my father's had joined the gold rush and he invited me to share his hut. Like my family, he came from the valley of the Doce River, a key region of my native state of Minas Gerais which also gave its name to the Companhia Vale do Rio Doce. What I didn't know was that word had spread among the *garimpeiros* that I might have been sent as a spy by the very company that claimed their mine. Certainly, with my long blond hair and moustache, I looked suspiciously out of place among the mud-splattered peons.

On my first day in Serra Pelada, as I observed this amazing human extravaganza, something unexpected happened. I began to hear the noise of shovels and pickaxes banging angrily against each other and I saw thousands of hostile faces looking up at me. When I climbed down into the hole, no one spoke to me; some even purposely soiled my khaki outfit with mud. It quickly became clear to me that I was the target of a spontaneous protest against the powerful CVRD mining company.

What followed was even stranger. A policeman approached me and shouted, "Gringo, your passport!" I answered that I had none, that I was a Brazilian, but he didn't believe me. "Come along," he yelled, handcuffing me and dragging me away as I desperately tried to protect my cameras. Suddenly, the prospectors' mood changed. Knowing that no government official would dare mistreat someone from CVRD, they began shouting abuse at the policeman.

When I reached the police post, the officer in charge also asked for my passport. "I'm a *mineiro*," I said, spelling out that I came from the state of Minas Gerais. And when I showed him my Brazilian identity card, he immediately apologized and released me. As I walked back to the edge of the mine, I was amazed to hear applause echoing up from the deep hole. I had been accepted. This run-in with the hated police would help me enormously in the weeks ahead.

I soon learned that what at first glance looked like a disorderly movement of men was in fact a highly sophisticated system, in which every one of the more than 50,000 men working there knew the role he had chosen to play. In broad terms, each operation involved three classes: the owner of the plot, usually a pioneer who received it as a concession from the local cooperative; the so-called *capitalista*, who invested in the operation; and 30, or so peons or digger/carriers.

These classes did not mirror the inequities of Brazil's social structure: Here, blacks and whites and men of every shade between worked shoulder-to-shoulder, with gold the only color on their minds. Further, just as the *capitalista* might simply be a peasant who sold his cattle and chose to invest in an owner's plot, the peon could easily be a university graduate, a truck driver, a town baker or a farmworker who imagined instant wealth and headed to Serra Pelada.

The hardest work was of course done by the peons: Their job was to dig earth from their assigned plots, fill sacks weighing 40 or so kilos and then carry them up steep slopes, along narrow paths and, finally, up ladders. To a man, their bodies, their feet, their faces, their necks were the color of ochre, stained by the iron ore in the earth they had excavated.

Most days I joined the workers in the hole, photographing them as they chopped away at their tiny flat plots, for the most part at different levels like a chaotically designed terraced hillside. Some plots were perilously perched as if on a cliff's edge. Sometimes a *capitalista*, standing out in his white hat and clean clothes, would come down to supervise. From below, I could also observe the heavily laden men as they worked their way upwards, so tightly packed that any hesitation or misstep could disrupt the entire flow, bringing shouts of alarm and loud curses. Then, a mere dozen meters away, with no ladders to guide them, those who had delivered their sacks were hurrying back to their plots, having learned that leaping from plot to plot was less risky than sliding down the steep muddy paths.

Inevitably, there were accidents, although fewer than you might expect. I witnessed one in which a man fell maybe 15 meters into the thick muddy water at the bottom of the hole. People rushed to help him, but he had swallowed so much toxic liquid that he was unconscious when he was pulled out. Alas, he did not survive. Still worse occurred a few days after I left Serra Pelada. Heavy rain provoked a landslide that buried ten men under tons of earth. In theory, work was meant to stop during the rainy season, but for many peons the thirst for gold was stronger than the fear for their lives.

Occasionally, a vein of gold was struck in a plot and cries of excitement echoed across the hole. Immediately, the diggers began feverishly shoveling earth into sacks, which were then piled up beside their plot. When the moment came to carry them up, the peons dressed for the occasion, putting on clean brightly colored striped shirts so that the guards could easily identify them and protect their cargo from theft. Having delivered the loads to their *capitalista*, each peon could pick one sack, with the rest kept by the boss. He then panned his chosen earth to discover if his golden day had finally come.

This was often the scene when I emerged from the hole: Scores of men on their haunches, swilling muddy water in broad pans until the more solid materials sank to the bottom, their eyes peeled for any suggestion of gold. I sensed a strange contradiction: These men were in a terrible hurry to get rich yet they needed immense patience in order to do so.

Early in the life of the mine, the government opened a branch of a state-owned bank, the Caixa Econômica Federal, which alone was authorized to buy gold, paying 15 percent less than the day's price on the London Metal Exchange. There was also a smelter where plot owners or peons could melt their gold into bars. And every day, giving hope to the luckless thousands still laboring in the mine, men brought their small fortunes to the bank to be exchanged for thick wads of paper money.

Discouragingly, perhaps as many as 99 percent of the sacks brought up each day contained nothing but earth and stones. And this detritus had to be dumped some way away to leave room for the next deliveries. Eager to keep their peons working from 7:00 am to 6:00 pm every day except Sunday, the *capitalista*s provided them with a small wage at the end of each day as well as with food and a shack or hut for shelter. They were of course free to walk away, to return home empty-handed, but few did. They had become slaves to gold.

A mood of latent violence prevailed and, while only the police were armed, fights would break out and these could escalate dangerously: One policeman who shot a worker was himself almost stoned to death by a mob. And on more than one occasion angry peons responded to mistreatment by seizing an abusive policeman and dragging him violently to the police post, where he would arrive bruised, battered, and repentant. A peculiar mutual disdain also fed tensions between police and peons: The police, who were earning a secure wage and working for the state government, considered themselves a rung above the mud-stained peons; in contrast, the workers "knew" that one day they would be far richer than any armed guard.

In a nearby settlement, along with a few stores selling sodas and bananas, an inevitable brothel offered some relief for this tight community of driven men. This in turn drew a group of evangelists to preach fruitlessly on Sundays against the sins of the flesh. Sometimes a solitary troubadour would appear, with an accordion to accompany his offering of melancholic songs about life in the Brazilian outback. There was also a homosexual "wing" to the housing area whose well-organized members also worked in the mine without apparent discrimination. In a sense, this was an equal opportunity employment since everyone shared the same chimera.

But gold is a fickle lover. While some fortunate prospectors left Serra Pelada with money, bought farms and businesses and never felt betrayed, others who found gold and believed more fortune awaited them eventually lost everything they had gained. My father's friend was like that. He found 97 kilograms of gold, reinvested his earnings in new plots and fresh teams of peons and eventually left the mine empty-handed.

Records show that by the time the concession was returned to the CVRD mining company in 1992, some 30 tons of gold had been extracted from Serra Pelada, with a value of some $400 million. They also show that output fell steadily each year, to as little as 245 kilos in 1990, representing a mere $3.5 million in sales. One reason was that the most accessible gold had already been removed. Another was that the deeper the digging, the more water entered the mine and hindered the work.

The belief that the mine still contains as much as 350 tons of gold, platinum, and palladium prompted a Canadian company, Colossus Minerals, to form a joint venture in 2011 with the surviving Serra Pelada Prospectors' Cooperative to dredge for its hidden wealth. Three years later, Brazil's Congress revoked the concession.

But the mania for gold does not easily wane. Some miners who abandoned Serra Pelada kept on chasing their dream hundreds of miles to the north in the Amazon jungle. And they were soon joined by many others who learned that gold could be found in the network of streams and rivers feeding the Amazon itself. By cutting down stretches of forest and poisoning rivers with mercury used for separating gold from mud, these desperate adventurers are seriously damaging the delicate ecology of the rainforest and disrupting the Indian way of life, even that of uncontacted tribes.

When their outposts are discovered, police or army units destroy their generators and pumps and drive them out of the jungle. But many soon return, willing to risk malaria, yellow fever, typhoid, snake bites, accidents, and arrest in the belief that one day fate will reward them. Sadly, even this is an illusion: When they do find gold, real wealth invariably goes to the middlemen who market the fruit of their suffering. Gold fever may keep them working, but for the most part it is also gold fever that keeps them poor.

As for Serra Pelada, the area has also returned to poverty, leaving a scarred landscape and a mighty lake 200 meters deep. For a decade, it evoked the long-promised El Dorado, but today Brazil's wildest gold rush is merely the stuff of legend, kept alive by a few happy memories, many pained regrets—and photographs.

Sebastião Salgado
Paris, 2019

GOLDFIEBER AM AMAZONAS

Als ich zum ersten Mal nach Serra Pelada kam, war ich sprachlos. Vor mir sah ich ein riesiges Loch mit einem Durchmesser von vielleicht 200 Metern und fast genauso tief. In ihm ein Gewimmel von mehreren zehntausend Männern, kaum bekleidet. Ungefähr die Hälfte von ihnen schleppte schwere Säcke über breite Holzleitern nach oben, während die anderen über schlammige Abhänge zurück in den Schlund der Grube rutschten.

Natürlich wusste ich, was sie dort taten. Sie suchten nach Gold. Und ich war zu dieser entlegenen Bergkuppe am Rande des Amazonas-Regenwaldes in Brasiliens nördlichem Bundesstaat Pará gereist, um sie zu fotografieren. Trotzdem war ich schlecht auf dieses außerordentliche Schauspiel vorbereitet.

Was hat dieses unspektakuläre gelbe Metall nur an sich, dass es die Menschen dazu bringt, ihre Heimat zu verlassen, all ihre Habe zu verkaufen und einen ganzen Kontinent zu durchqueren, um ihr Leben, ihre Knochen und ihre Gesundheit für einen Traum aufs Spiel zu setzen? Serra Pelada – portugiesisch der „kahle Berg" – war nicht der erste Ort in Nord- oder Südamerika, der ein derartiges Goldfieber auslöste. Beinahe fünf Jahrhunderte zuvor führte die Suche nach dem El Dorado zur Eroberung eines ganzen Kontinents. Nachdem Hernán Cortés 1519 an der Küste Mexikos gelandet war, sagte er dem Herrscher der Azteken, dass seine Männer an einer Krankheit litten, die nur Gold heilen könne. Und die Gier nach schnellem Reichtum erreichte im kalifornischen *Gold Rush* der 1840er-Jahre und in der wilden Goldsuche am kanadischen Klondike River 50 Jahre später neue Höhen des Wahnsinns.

Und wie die gierigen Goldsucher früherer Zeiten waren auch die *garimpeiros* in Serra Pelada von derselben alles verzehrenden Leidenschaft angesteckt, schnell reich zu werden.

1979 wurde dort in einem Fluss Gold entdeckt, und innerhalb weniger Wochen begann der Goldrausch. Die ersten Männer, die ankamen, bildeten eine Kooperative, die den Pionieren das Schürfrecht für 2 x 3 Meter große Parzellen übertrug. Ihnen folgten bald Zehntausende, die als Tagelöhner das Graben übernahmen und die Erde zu den Parzellenbesitzern und ihren Mitinvestoren schleppten.

Um die Ordnung aufrechtzuerhalten, entsandte die brasilianische Militärregierung die Bundespolizei nach Serra Pelada. Sie erlaubte, dass dort weitergearbeitet wurde, verbot aber

Schusswaffen, Alkohol und Frauen. Offiziell lag das Schürfrecht bei dem staatlichen Bergbaugiganten Companhia Vale do Rio Doce (CVRD, heute in Vale umbenannt), aber es war völlig aussichtslos, den Wahnsinn des ungeregelten Abbaus stoppen zu wollen, der dort in fast biblischen Dimensionen stattfand.

Ich hatte bereits 1980 versucht, Serra Pelada zu besuchen, war aber durch die Gesetze daran gehindert worden, mit denen das brasilianische Militär, nachdem es 1964 an die Macht gekommen war, die Pressefreiheit und andere Grundrechte einschränkte. Es war auch wenig hilfreich, dass ich, um gegen die Diktatur zu protestieren, nach Europa ins Exil gegangen war. So hatte man mir für die nächsten fünf Jahre jeglichen Besuch der Mine verboten. Als das Militär jedoch nicht länger die Macht ausübte und die Staatspolizei von Pará für die Sicherheit zuständig war, erteilte mir die Kooperative der Goldsucher im September 1986 die Erlaubnis, dorthin zu fahren.

Durch einen glücklichen Zufall war auch ein Freund meines Vaters beim *Gold Rush* dabei und lud mich in seine Hütte ein. Wie meine Familie kam er aus dem Tal des Flusses Doce im Zentrum von Minas Gerais, wo ich geboren wurde und von dem die Companhia Vale do Rio Doce ihren Namen ableitete. Jedoch wusste ich nicht, dass sich unter den *garimpeiros* das Gerücht verbreitet hatte, die Bergbaugesellschaft, die ihre Mine beanspruchte, habe mich als Spion geschickt. Sicherlich sah ich unter den schmutzstarrenden Tagelöhnern mit meinen langen blonden Haaren und dem Schnurrbart verdächtig fremd aus.

An meinem ersten Tag in Serra Pelada, als ich dieses erstaunliche menschliche Schauspiel beobachtete, geschah etwas Außergewöhnliches. Ich hörte, wie Schaufeln und Spitzhacken heftig aufeinandergeschlagen wurden, und sah, dass Tausende von feindseligen Gesichtern zu mir hochschauten. Als ich in die Grube herunterstieg, sprach niemand mit mir, einige beschmutzten sogar mit Absicht meinen Khaki-Anzug mit Schlamm. Schnell wurde mir klar, dass ich zum Ziel eines spontanen Protestes gegen die mächtige CVRD-Bergbaugesellschaft geworden war.

Was dann folgte, war noch seltsamer. Ein Polizist näherte sich mir und rief: „Gringo, Deinen Pass!“ Ich antwortete ihm, dass ich keinen hätte und dass ich Brasilianer sei, aber er glaubte mir nicht. „Mitkommen!“, befahl er, legte mir Handschellen an und zog mich mit sich fort, während ich verzweifelt versuchte, meine Kameras zu schützen. Plötzlich änderte sich die Stimmung der Goldsucher. Da sie wussten, dass kein Beamter sich erlauben würde, einen Angestellten der CVRD so zu behandeln, überschütteten sie den Polizisten mit Beleidigungen.

Als wir die Polizeistation erreicht hatten, wollte der diensthabende Beamte ebenfalls meinen Pass sehen. „Ich bin ein *mineiro*“, sagte ich und erklärte ihm, dass ich aus Minas Gerais käme. Und als ich ihm meinen brasilianischen Personalausweis zeigte, entschuldigte er sich unverzüglich und ließ mich frei. Als ich zurück zum Rand der Grube ging, war ich verblüfft, aus der Tiefe Beifall zu hören. Ich wurde akzeptiert. Dieser Zusammenstoß mit der verhassten Polizei sollte mir in den folgenden Wochen noch sehr zugutekommen.

Bald stellte ich fest, dass das, was auf den ersten Blick wie ein chaotisches Menschengewimmel aussah, in Wahrheit ein hochkomplexes System war, in dem jeder der mehr als 50 000 dort arbeitenden Männer die Rolle kannte, die er dort zu spielen hatte. Ganz allgemein gesagt, waren bei jedem Arbeitsablauf drei „Klassen“ involviert: der Eigentümer des Claims, meist ein Pionier, der das Schürfrecht von der örtlichen Kooperative erhalten hatte, der sogenannte *capitalista*, der in das Unternehmen investierte, und für das Graben und Tragen etwa dreißig Tagelöhner.

Diese Gruppen waren kein Spiegel der Ungleichheiten brasilianischen Sozialstruktur: Hier arbeiteten Schwarze wie Weiße und Männer jeder anderen Hautfarbe Schulter an Schulter. Gold war die einzige Farbe, die sie im Sinn hatten. Und so wie der *capitalista* einfach ein Bauer sein konnte, der sein Vieh verkauft hatte, um das Geld in einen Claim zu investieren, konnte der Tagelöhner ebenso ein Universitätsabsolvent, ein Lastwagenfahrer, ein Bäcker aus der Stadt oder ein Farmarbeiter sein, der sich in der Hoffnung, schnell reich zu werden, nach Serra Pelada aufgemacht hatte.

Den härtesten Job erledigten natürlich die Tagelöhner: Ihre Aufgabe bestand darin, die Erde der zugewiesenen Parzellen abzutragen, sie in Säcke zu füllen, die ungefähr 40 Kilo wogen, und diese über steile Abhänge, entlang schmaler Pfade und am Ende die Leitern empor nach oben zu tragen. Bei allen von ihnen waren die Körper, Füße, Gesichter und Nacken ockerfarben vom Eisenerz in der Erde, die sie ausgehoben hatten.

An den meisten Tagen hielt ich mich bei den Arbeitern in der Grube auf und fotografierte sie, während sie die Erde ihrer winzigen, flachen Parzellen aufhackten, oft auf ganz unterschiedlichen Niveaus, sodass der terrassierte Hügel völlig chaotisch aussah. Einige Parzellen ragten gefährlich hervor, als lägen sie auf einer Klippe. Manchmal kam ein *capitalista*, unübersehbar mit seinem weißen Hut und der sauberen Kleidung, zur Kontrolle nach unten. Von dort konnte ich die schwer beladenen Männer auch dabei beobachten, wie sie sich nach oben vorarbeiteten, so dicht hintereinander gehend, dass jedes Zögern und jeder Fehltritt die ganze Kolonne durcheinanderbringen konnte, was laute Alarmrufe und Flüche zur Folge hatte. Dann, nur ein Dutzend Meter entfernt, eilten diejenigen, die ihre Säcke abgeliefert hatten, wieder zurück zu ihren Parzellen. Dabei sprangen sie, ohne Leitern als Hilfsmittel, von Absatz zu Absatz, wussten sie doch aus Erfahrung, dass dies weniger riskant war, als die steilen, schlammigen Pfade herunterzurutschen.

Natürlich gab es Unfälle, indes weniger als man denken könnte. Ich wurde Zeuge eines Unglücks, bei dem ein Mann etwa 15 Meter tief in das trübe, schlammige Wasser am Boden der Grube fiel. Die Arbeiter beeilten sich, ihm zu helfen, aber er hatte bereits so viel von dem giftigen Wasser geschluckt, dass er bewusstlos war, als man ihn herauszog. Traurigerweise überlebte er nicht. Noch Schlimmeres ereignete sich einige Tage später, nachdem ich Serra Pelada verlassen hatte. Ein heftiger Regen verursachte einen Erdrutsch und begrub zehn Männer unter Tonnen von Erde. Theoretisch sollte die Arbeit während der Regenzeit eingestellt werden, aber für viele Arbeiter war der Hunger nach Gold stärker als die Angst um das eigene Leben.

Gelegentlich entdeckte man eine Goldader auf einem Claim; dann wurden in der Grube aufgeregte Schreie laut. Unmittelbar danach begannen die Goldgräber, in fieberhafter Eile Erde in die Säcke zu schaufeln, die sie dann neben ihrem Claim auftürmten. Wenn es an der Zeit war, sie nach oben zu tragen, zogen die Arbeiter saubere, gestreifte Hemden an, sodass die Wächter sie leicht erkennen und ihre Lieferung vor Diebstahl schützen konnten. Wenn sie die Säcke bei ihrem *capitalista* abgeliefert hatten, durfte jeder Tagelöhner einen behalten, der Rest gehörte dem Chef. Der Arbeiter durchsuchte dann seinen Anteil in der Hoffnung, dass sein goldener Tag nun endlich gekommen sei.

Diese Szene sah ich oft, wenn ich aus der Grube auftauchte: Männer in der Hocke, die ihre großen Pfannen mit schmutzigem Wasser durchspülten, bis der festere Stoff zu Boden sank, während ihre Augen nach der kleinsten Andeutung von Gold forschten. Für mich ergab das einen seltsamen Widerspruch: Die Männer hatten es furchtbar eilig, reich zu werden, und wurden dabei zu einer unendlichen Geduld gezwungen.

Schon früh in der Geschichte dieser Mine hatte die Regierung dort die Filiale einer staatlichen Bank eröffnet, die Caixa Econômica Federal. Sie allein hatte das Recht, Gold zu kaufen, und zwar zu einem Preis 15 Prozent unter dem Tagespreis der Londoner Metallbörse. Es gab auch einen Ofen, in dem die Besitzer der Claims oder die Arbeiter ihr Gold zu Barren einschmelzen konnten. Jeden Tag brachten Männer ihre kleinen Vermögen zur Bank und erhielten dafür dicke Bündel Papiergeld, was den glücklosen Tausenden, die noch in der Mine arbeiteten, Hoffnung machte.

Entmutigenderweise enthielten beinahe 99 Prozent der Säcke, die jeden Tag nach oben gebracht wurden, nichts als Erde und Steine. Und dieser Schutt musste irgendwie entsorgt werden, um für die nächsten Lieferungen Platz zu schaffen. Eifrig darauf bedacht, dass ihre Tagelöhner jeden Tag außer am Sonntag von 7 Uhr morgens bis 6 Uhr abends schufteten, gaben ihnen die *capitalistas* am Ende jeden Tages einen geringen Lohn und versorgten sie mit Lebensmitteln und einer Hütte oder Baracke. Natürlich konnten sie jederzeit gehen und mit leeren Händen nach Hause zurückkehren, aber nur wenige taten das. Sie waren zu Sklaven des Goldes geworden.

Eine latente Aggressivität herrschte vor, und obwohl nur die Polizei bewaffnet war, brachen Schlägereien aus, die gefährlich eskalieren konnten: Ein Polizist, der einen Arbeiter erschossen hatte, wurde vom Mob fast zu Tode gesteinigt. Und mehr als einmal reagierten wütende Arbeiter auf die schlechte Behandlung durch einen Polizisten, indem sie diesen packten und gewaltsam zur Polizeistation schleiften, wo er dann geprügelt, verletzt und reumütig ankam. Es gab noch eine besondere Form von gegenseitiger Abneigung, die ebenfalls zu Spannungen zwischen der Polizei und den Arbeitern führte: Die Polizisten bezogen ein sicheres Gehalt und arbeiteten für die Regierung. Sie hielten sich gegenüber den schmutzbedeckten Tagelöhnern für überlegen. Doch diese „wussten", dass sie eines Tages reicher sein würden als jeder der bewaffneten Ordnungshüter.

In einer nahegelegenen Siedlung gab es einige Läden, die Getränke und Bananen verkauften, und natürlich das unvermeidliche Bordell, das den von der Goldgier getriebenen Männern dieser engen Gemeinschaft ein wenig Amüsement bot. Dies wiederum zog eine Gruppe von Missionaren an, die jeden Sonntag erfolglos gegen die Sünden des Fleisches predigten. Manchmal kam ein einsamer Troubadour des Weges, der seine melancholischen Lieder über das Leben im brasilianischen Hinterland auf einem Akkordeon begleitete. Es gab außerdem einen homosexuellen „Flügel", dessen gut organisierte Mitglieder ebenfalls in der Mine arbeiteten, ohne dass man sie diskriminiert hätte. In gewisser Weise war dies hier die Arbeit von Gleichgestellten, weil jeder derselben Schimäre hinterherjagte.

Aber das Gold ist ein wankelmütiger Liebhaber. Während einige glückliche Goldgräber Serra Pelada mit Geld verließen, eine Farm oder ein Geschäft kauften und zufrieden waren, glaubten andere, die Gold gefunden hatten, sie könnten noch reicher werden und verloren am Ende alles, was sie zusammengebracht hatten. So erging es dem Freund meines Vaters. Er fand 97 Kilo Gold, investierte seinen Gewinn in weitere Parzellen und neue Mannschaften von Tagelöhnern und verließ die Mine am Ende mit leeren Händen.

Die offiziellen Dokumente belegen, dass man zu der Zeit, als die Konzession 1992 an die Bergbaugesellschaft CVRD zurückgegeben wurde, aus der Mine in Serra Pelada etwa 30

Tonnen Gold im Wert von rund 400 Millionen Dollar gewonnen hatte. Sie zeigen auch, dass die Ausbeute Jahr für Jahr weniger wurde, bis hin zu nur 245 Kilo im Jahr 1990, was nur noch einem Verkaufswert von etwa 3,5 Millionen Dollar entsprach. Ein Grund dafür war, dass das leicht zugängliche Gold bereits gefördert worden war. Ein anderer, dass je tiefer man grub, desto mehr Wasser in die Grube floss, was die Arbeit erschwerte.

Die Vermutung, dass die Mine noch weitere 350 Tonnen Gold, Platin und Palladium enthielt, brachte die kanadische Firma Colossus Minerals dazu, im Jahr 2011 ein Joint Venture mit der noch existierenden Goldsucher-Kooperative von Serra Pelada einzugehen, um nach diesem verborgenen Schatz zu graben. Drei Jahre später kündigte der brasilianische Kongress die Konzession.

Aber die Gier nach Gold ist nicht so leicht zu stillen. Einige Minenarbeiter, die Serra Pelada verlassen hatten, jagten ihrem Traum Hunderte von Kilometern weiter nördlich im Amazonas-Dschungel weiter nach. Und bald folgten ihnen andere, die erfahren hatten, dass man Gold in den Bächen und Flüssen finden könne, die in den Amazonas münden. Dort holzen sie ganze Waldgebiet ab und vergiften die Flüsse mit Quecksilber, um das Gold aus dem Schlamm zu waschen. So schädigen diese hemmungslosen Abenteurer das sensible Ökosystem des Regenwaldes schwer und die Lebensgrundlage der eng mit der Natur verbundenen Ureinwohner, sogar bislang isoliert lebender Stämme.

Wenn solche illegalen Grabungen entdeckt werden, zerstören die Polizei oder Armeeeinheiten ihre Generatoren und Pumpen und vertreiben sie aus dem Dschungel. Aber viele von ihnen kehren bald wieder zurück. Sie riskieren Malaria, Gelbfieber, Typhus, Schlangenbisse, Unfälle und Gefängnis im Glauben, dass ihnen das Glück eines Tages hold sein wird. Traurigerweise ist auch das eine Illusion. Wenn sie Gold finden, geht der Löwenanteil nach wie vor an die Zwischenhändler, die die Früchte ihres Leidens vermarkten. Das Goldfieber mag sie zur Arbeit antreiben, aber meistens sorgt es auch dafür, dass sie arm bleiben.

Serra Pelada ist heute wieder eine arme Region. Geblieben ist eine Landschaft voller Narben und ein riesiger, 200 Meter tiefer See. Ein Jahrzehnt lang sah es dort wie das erträumte El Dorado aus, aber heute ist Brasiliens wildester Goldrausch nur noch Stoff für Legenden. Am Leben erhalten werden diese durch ein paar glückliche und viele schmerzliche Erinnerungen – und durch Fotografien.

Sebastião Salgado
Paris, 2019

L'OR : UNE FIÈVRE AMAZONIENNE

La première fois que je vis la mine de Serra Pelada, je restai sans voix. Devant moi s'ouvrait une immense fosse de presque deux cents mètres de diamètre pour une profondeur identique, où grouillaient des dizaines de milliers d'hommes à peine vêtus, dont la moitié montait de lourds sacs de terre sur de grandes échelles en bois, les autres dévalant les pentes boueuses pour rejoindre le gouffre.

Bien sûr, je savais ce qu'ils faisaient. Ils cherchaient de l'or. Et moi, j'avais roulé jusqu'à cette colline perdue en bordure de la forêt pluviale amazonienne, dans l'État du Pará, au nord du Brésil, pour les photographier. Je le savais, et pourtant j'étais mal préparé pour appréhender cet extraordinaire spectacle.

Qu'est-ce qui, dans ce métal jaune et sans éclat, pousse des hommes à abandonner leur foyer, vendre leurs biens et traverser un continent pour risquer leur vie, leurs membres et leur santé mentale au nom d'un rêve ? Ce n'est pas à Serra Pelada – la montagne pelée en portugais – que cette fièvre fatale a frappé les Amériques pour la première fois. Presque cinq siècles plus tôt, la quête de l'Eldorado avait abouti à la conquête de tout un continent. Après avoir débarqué sur la côte du Mexique en 1519, Hernán Cortès avait expliqué à l'empereur aztèque que ses hommes souffraient d'une maladie qui ne pouvait être guérie que par l'or. Ce désir pressant de s'enrichir aussi vite que possible perdura, atteignant des sommets de folie lors de la ruée vers l'or des années 1840 en Californie, et dans le Klondike canadien cinquante ans plus tard.

Comme les premières vagues de spéculateurs avant eux, au regard fou, les prospecteurs ou garimpeiros *de Serra Pelada étaient contaminés par une ferveur dévorante.*

L'or fut découvert dans un ruisseau de la région en 1979 et, en quelques semaines, une nouvelle ruée vers l'or avait débuté. Les premiers arrivés formèrent une coopérative qui assigna à chaque pionnier une parcelle de terrain de 2 x 3 mètres. Des milliers d'hommes affluèrent bientôt – les péons –, afin de creuser et transporter la terre censée contenir le fameux métal pour le compte des propriétaires de parcelles et de leurs co-investisseurs.

Pour maintenir l'ordre, le gouvernement des militaires brésiliens envoya la police fédérale sur le site, autorisant la poursuite de l'exploitation mais interdisant les armes, l'alcool et les femmes. Officiellement, le droit d'exploiter le terrain avait été concédé à un géant minier possédé par l'État, la Companhia Vale do Rio Doce (CVRD), aujourd'hui renommée Vale, mais il était impossible de mettre un terme à l'exploitation sauvage et frénétique qui se déroulait maintenant dans des proportions quasi bibliques.

J'avais déjà essayé de me rendre à Serra Pelada en 1980 mais en avais été empêché par les multiples restrictions de liberté imposées par les militaires au pouvoir depuis 1964, visant notamment la presse. Mon choix de m'exiler en Europe pour protester contre cette dictature n'avait évidemment pas arrangé les choses. Pendant cinq ans, on me refusa l'accès à la mine. Enfin, en septembre 1986, après que les militaires eurent quitté le pouvoir, la police de l'État du Pará fut chargée de la sécurité et la coopérative des prospecteurs me donna son feu vert.

Par chance, un ami de mon père s'était joint à cette ruée vers l'or et m'invita à partager sa hutte. Comme ma famille, il venait de la vallée du Doce, région clé de mon État natal du Minas Gerais, qui avait donné son nom à la Companhia Vale do Rio Doce. Ce que je ne savais pas, c'est que l'idée s'était répandue parmi les garimpeiros *que j'étais peut-être un espion envoyé par la CVRD désireuse de reprendre le contrôle de la mine. Avec ma moustache et mes longs cheveux blonds, je devais paraître suspicieusement déplacé au milieu des péons couverts de boue.*

Le jour de mon arrivée à Serra Pelada, alors que j'observais cette extravagante manifestation de la folie humaine, quelque chose de surprenant se produisit. J'entendis d'abord le bruit de pelles et de pioches violemment entrechoquées et je vis des milliers de visages hostiles se tourner vers moi. Lorsque je descendis dans le trou, personne ne m'adressa la parole et certains salirent volontairement ma tenue kaki avec de la boue. Je compris rapidement que j'étais la cible d'une protestation spontanée contre la puissante société minière.

La suite fut plus étrange encore. Un policier s'approcha en me criant dessus : « Gringo, ton passeport ! » Je répondis que je n'en avais pas, que j'étais brésilien, mais il ne me crut pas. « Allez viens ! » me hurla-t-il, en me passant les menottes et me tirant derrière lui alors que j'essayais désespérément de protéger mes appareils. Brusquement, l'attitude des prospecteurs changea. Sachant qu'aucun personnage officiel n'oserait maltraiter quelqu'un de la CVRD, ils commencèrent à insulter le policier.

À mon arrivée au poste de police, un responsable me demanda lui aussi mon passeport. « Je suis un Mineiro *», dis-je pour l'informer que je venais du Minas Gerais. Puis je lui montrai ma carte d'identité brésilienne. Il s'excusa immédiatement et me laissa partir. De retour à la mine, je fus étonné d'être reçu par des applaudissements montant des profondeurs de la fosse. J'avais été accepté. Cet accrochage avec une police détestée allait m'aider énormément dans les semaines à venir.*

Je compris rapidement que ce qui ressemblait au premier regard à un mouvement désordonné était en fait un système hautement sophistiqué, dans lequel chacun des 50 000 hommes y travaillant connaissait son rôle. Pour résumer, chaque opération impliquait trois catégories de participants : le propriétaire de la parcelle, généralement un pionnier qui l'avait reçue en concession de la coopérative locale, le surnommé capitalista *qui avait investi dans l'opération et une trentaine de péons, qui creusaient et transportaient la terre.*

Ces catégories ne reflétaient pas les inégalités de la structure sociale brésilienne. Ici, les hommes blancs, noirs et de toutes les nuances entre les deux travaillaient épaule contre épaule, l'or étant la seule couleur valable à leurs yeux. Par ailleurs, de même que le capitalista *pouvait n'être qu'un paysan qui avait vendu son troupeau et décidé d'investir dans la parcelle d'un propriétaire, le péon pouvait aisément être un diplômé d'université, un conducteur de camion, un boulanger de la ville ou un ouvrier agricole, qui, rêvant d'une richesse instantanée, avait pris la route de Serra Pelada.*

Le travail le plus dur était, bien sûr, laissé aux péons. Leur fonction consistait à creuser le sol de leur parcelle, à remplir des sacs de terre de 40 kilos environ, à les remonter le long d'étroits chemins escarpés et enfin à grimper aux échelles. Tous sans exception – leur corps, leurs pieds, leur visage, leur cou – étaient de la couleur ocre de la terre qu'ils prélevaient, colorée par le minerai de fer.

La plupart du temps, je rejoignais les ouvriers dans le trou pour les photographier en train de travailler leurs petites parcelles, presque toutes situées à des niveaux différents, un paysage de terrasses cultivées aménagées n'importe comment. Certaines étaient dangereusement perchées, comme au bord d'une falaise. Parfois, un capitalista, *debout, chapeau blanc et vêtements propres, descendait pour superviser le travail. D'en bas, je pouvais aussi observer les hommes lourdement chargés remonter la pente, si serrés les uns contre les autres que la moindre hésitation, le moindre faux pas, pouvait rompre leur flux incessant, provoquant cris d'alerte et chapelets de jurons. Puis, une douzaine de mètres plus loin, sans échelle pour les aider, ceux qui avaient livré leurs sacs se dépêchaient de redescendre à leur parcelle en sautant de l'une à l'autre, une méthode moins risquée que de glisser sur les sentiers boueux en pente raide.*

Inévitablement, des accidents se produisaient, bien que moins nombreux que l'on aurait pu le penser. Je fus témoin d'une chute de presque quinze mètres d'un homme dans l'eau boueuse des profondeurs de la fosse. Des péons se précipitèrent pour l'aider mais il avait avalé tellement de liquide toxique qu'il était inconscient lorsqu'il fut extirpé. Malheureusement, il n'a pas survécu. Un accident bien pire se produisit quelques jours après mon départ de Serra Pelada, quand une forte pluie provoqua un glissement de terrain, engloutissant dix hommes sous des tonnes de terre.

En théorie, ils devaient s'arrêter de travailler pendant la saison des pluies, mais pour de nombreux péons la soif de l'or était plus forte que la crainte de la mort.

Parfois, une veine d'or était découverte dans une parcelle et des cris d'excitation résonnaient en écho dans le site. Immédiatement, les hommes chargés de creuser se mettaient à remplir fiévreusement les sacs de terre qu'ils entassaient à proximité. Au moment de les remonter, les péons revêtaient pour l'occasion des chemises bariolées propres, afin que les gardes puissent les identifier aisément et protéger leur chargement des voleurs. Après avoir livré son capitalista, *chaque péon avait droit à un sac, le reste étant conservé par le patron. Il lavait ensuite son morceau de terre à la battée en espérant que son jour de chance était enfin arrivé.*

Quand je remontais de la fosse, je voyais souvent la même scène : des centaines d'hommes accroupis faisant tournoyer l'eau boueuse dans de grandes battées jusqu'à ce que les éléments les plus solides coulent au fond, les yeux grands ouverts pour ne pas rater la moindre trace d'or. J'y voyais une étrange contradiction. Ces hommes étaient terriblement pressés de devenir riches et avaient pourtant besoin d'une immense patience pour y arriver.

Au tout début de l'histoire de la mine, les autorités avaient ouvert une succursale de la Caixa Econômica Federal, une banque nationale, seule autorisée à acheter l'or, à 15 % de moins que le cours officiel à la Bourse spécialisée de Londres. Il y avait aussi une fonderie où les propriétaires de parcelles comme les péons pouvaient transformer leur or en barres. Chaque jour, alimentant l'espoir de milliers de personnes, moins chanceuses, qui travaillaient encore à la mine, des hommes apportaient ainsi leur petite fortune à la banque en échange d'épaisses liasses de billets.

Malheureusement, peut-être 99 % des sacs remontés à la surface chaque jour ne contenaient rien d'autre que des cailloux et de la terre – et ces déchets devaient être jetés assez loin pour laisser place aux livraisons à venir. Soucieux de faire travailler leurs péons de 7 heures à 18 heures chaque jour sauf le dimanche, les capitalistas *leur offraient un petit salaire à la fin de chaque journée ainsi que de la nourriture et une cabane ou une hutte pour s'abriter. Ils étaient libres de s'en aller et de retourner chez eux les mains vides, mais peu choisissaient de le faire. Ils étaient devenus les esclaves de l'or.*

Sur place prévalait une atmosphère de violence latente et, bien que seule la police fût armée, des bagarres éclataient et pouvaient connaître des escalades dangereuses. Un policier qui avait tiré sur un travailleur fut presque lapidé à mort par la foule. Plus d'une fois, des péons en colère répondirent aux mauvais traitements en s'emparant d'un policier avant de le traîner au poste de police par la force, où il arrivait contusionné, battu et repentant. Un dédain mutuel alimentait les tensions entre la police et les péons. Les policiers, qui percevaient un salaire chaque mois et travaillaient pour l'administration de l'État, se considéraient au-dessus de ces péons couverts de boue. Par contraste, les travailleurs « savaient » qu'un jour ils seraient bien plus riches que n'importe lequel de ces gardes armés.

Dans un campement voisin, parmi quelques boutiques proposant sodas et bananes, l'inévitable bordel offrait un répit à cette communauté d'hommes déterminés. Cela poussa ensuite un groupe d'évangélistes à prêcher le dimanche contre les péchés de la chair, en vain. Parfois arrivait un chanteur ambulant solitaire, qui s'accompagnait à l'accordéon et entonnait des chansons mélancoliques sur la vie dans l'arrière-pays brésilien. Il y avait aussi une « aile » homosexuelle dans le campement dont les membres, bien organisés, travaillaient à la mine sans discrimination apparente. En un sens, il y avait là une certaine équité dans l'accès à l'emploi, puisque tout le monde partageait les mêmes chimères.

Mais l'or est un amant inconstant. Si certains prospecteurs chanceux quittèrent Serra Pelada les poches pleines, achetèrent des fermes et des entreprises sans jamais se sentir trompés, d'autres, qui avaient découvert de l'or et croyaient pouvoir en trouver davantage, perdirent tout ce qu'ils avaient gagné. Ce fut le cas pour l'ami de mon père. Il trouva 97 kilos d'or, réinvestit son argent dans de nouvelles parcelles et de nouvelles équipes de péons et finit par quitter la mine les mains vides.

Les archives montrent que, lorsque la concession revint à la CVRD en 1992, quelque 30 tonnes d'or avaient été extraites de Serra Pelada pour une valeur d'environ 400 millions de dollars. Elles montrent aussi que la production chuta régulièrement chaque année, jusqu'à 245 kilos en 1990, représentant tout au plus 3,5 millions de dollars. Une des raisons en était que l'or le plus accessible avait déjà été extrait. Par ailleurs, plus le trou se faisait profond, plus l'eau rentrait dans la mine et gênait le travail de prospection.

Pensant que la mine contenait encore près de 350 tonnes d'or, de platine et de palladium, une société canadienne, Colossus Minerals, constitua en 2011 une coentreprise avec ce qui restait de la coopérative des prospecteurs de Serra Pelada pour tenter de dénicher ce trésor caché. Trois ans plus tard, le Congrès brésilien révoqua la concession.

Mais la fièvre de l'or ne disparaît pas aussi facilement. Certains mineurs qui avaient abandonné Serra Pelada ont poursuivi leur rêve à des centaines de kilomètres vers le nord, dans la jungle amazonienne. Ils furent bientôt rejoints par beaucoup d'autres qui avaient appris que le précieux métal pouvait être trouvé dans le réseau de ruisseaux et de rivières qui alimentent l'Amazone. En abattant des parties de la forêt et en empoisonnant les rivières avec du mercure, utilisé pour séparer l'or de la boue, ces aventuriers désespérés portent sérieusement atteinte à l'écologie délicate de la forêt pluviale et compromettent le style de vie des Indiens, dont celui de tribus non contactées.

Lorsque leurs campements sont découverts, la police ou des unités de l'armée détruisent les générateurs et les pompes et chassent les mineurs de la jungle. Mais beaucoup y retournent aussitôt, prêts à risquer la malaria, la fièvre jaune, la typhoïde, les morsures de serpents, les accidents et les arrestations en espérant qu'un jour le destin les récompensera. Tristement, il s'agit là encore d'une illusion. Lorsqu'ils trouvent de l'or, l'essentiel du gain va immanquablement aux intermédiaires qui commercialisent le fruit de leurs souffrances. La fièvre de l'or leur donne peut-être du travail mais les maintient presque tous dans la misère.

À Serra Pelada, la zone aussi est retombée dans la pauvreté. La mine n'est plus qu'un paysage traumatisé autour d'un lac de deux cents mètres de profondeur. Pendant une décennie, il a pu évoquer l'Eldorado tant attendu, mais aujourd'hui, la ruée sauvage vers l'or du Brésil n'est plus que l'étoffe de légendes, maintenue en vie par quelques souvenirs heureux, beaucoup de regrets douloureux et des photographies.

Sebastião Salgado
Paris, 2019

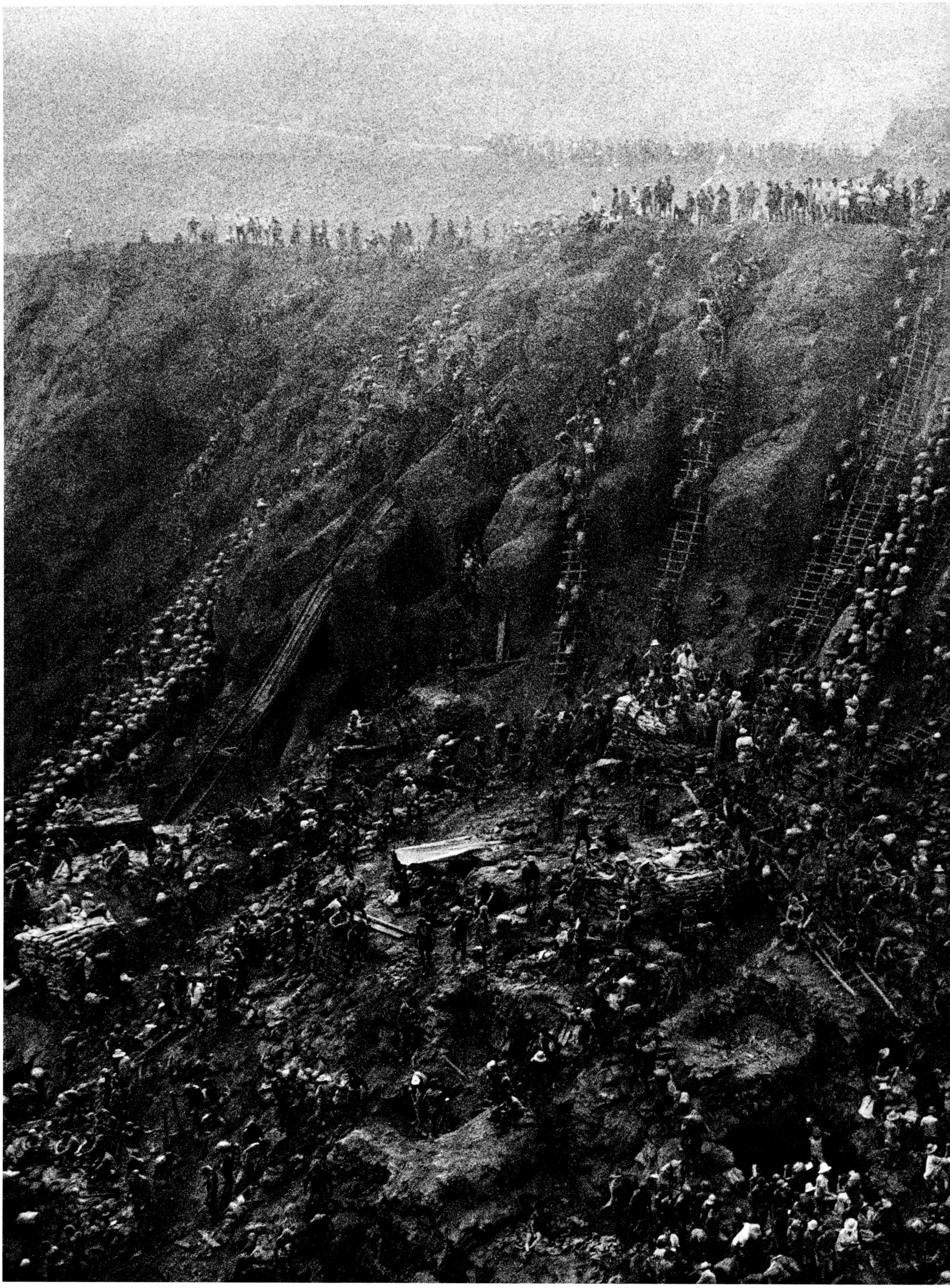

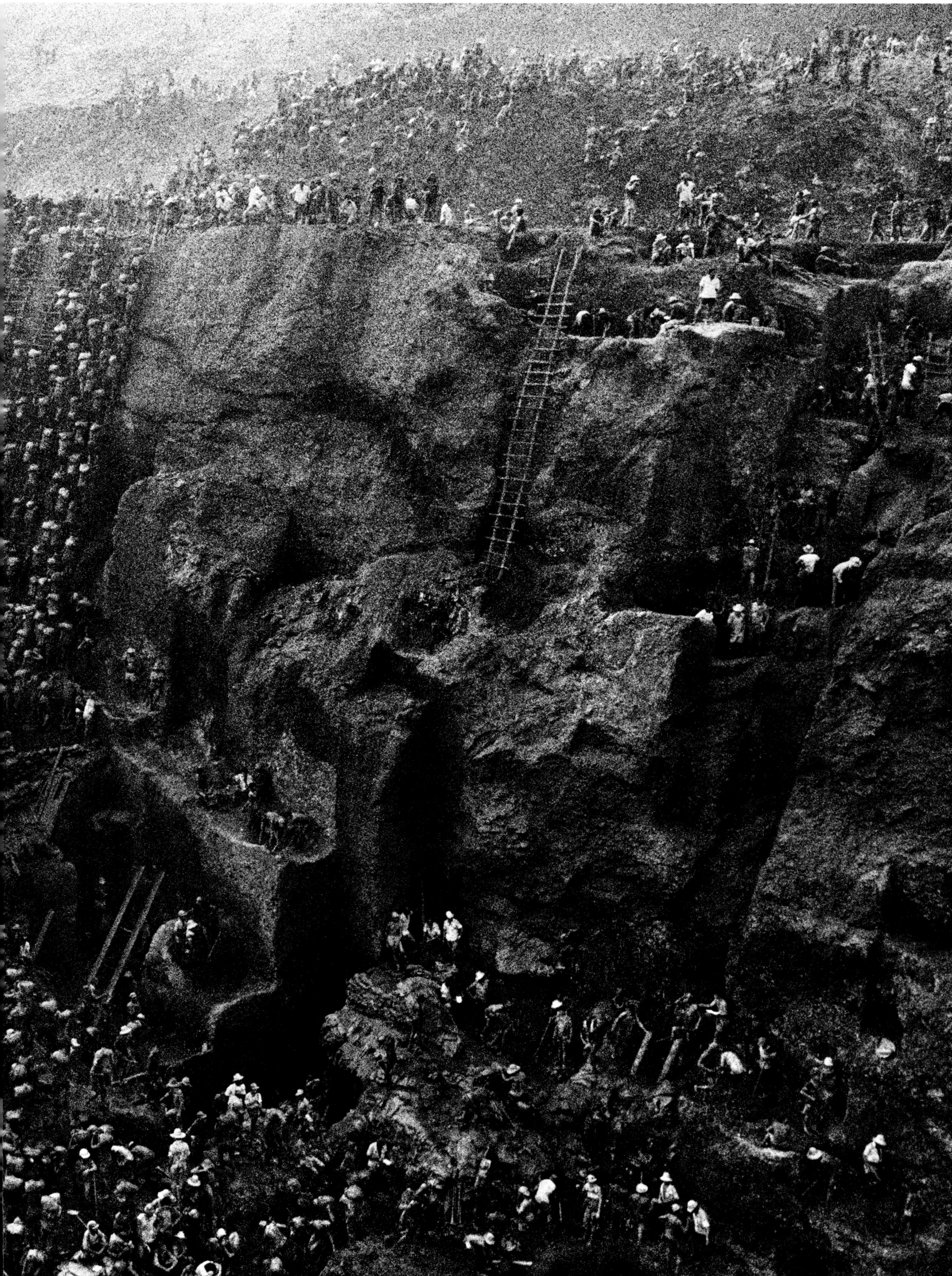

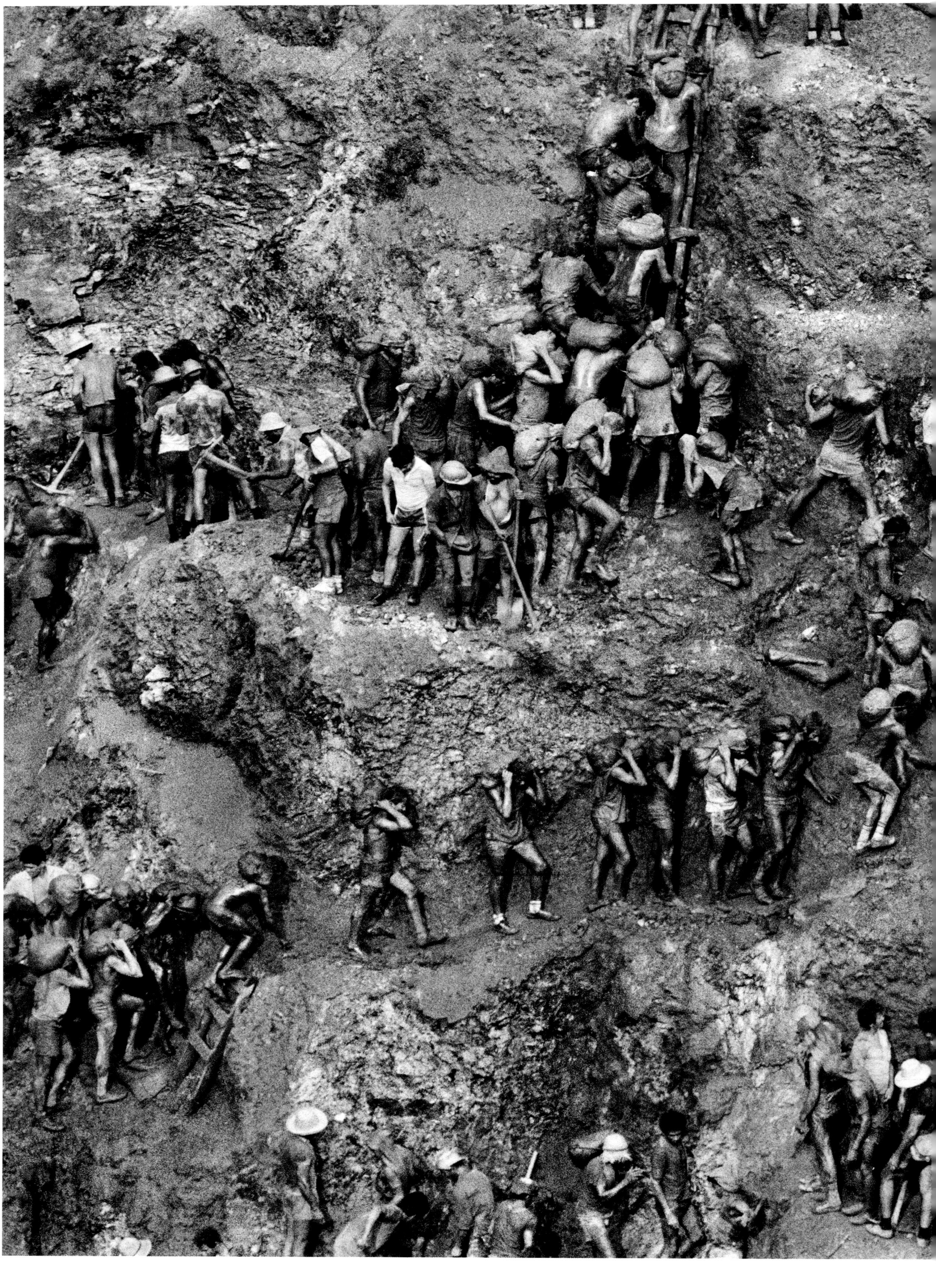

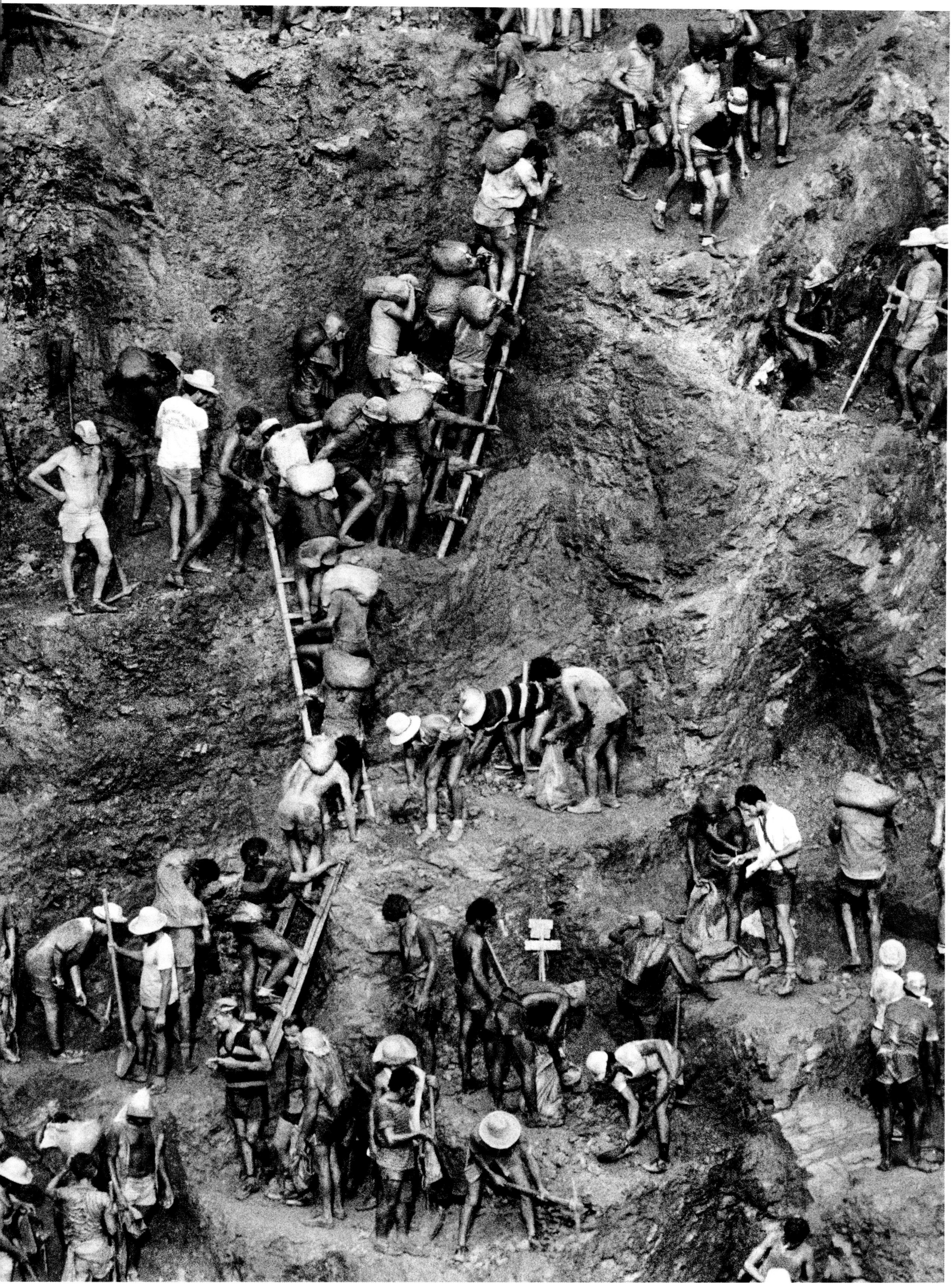

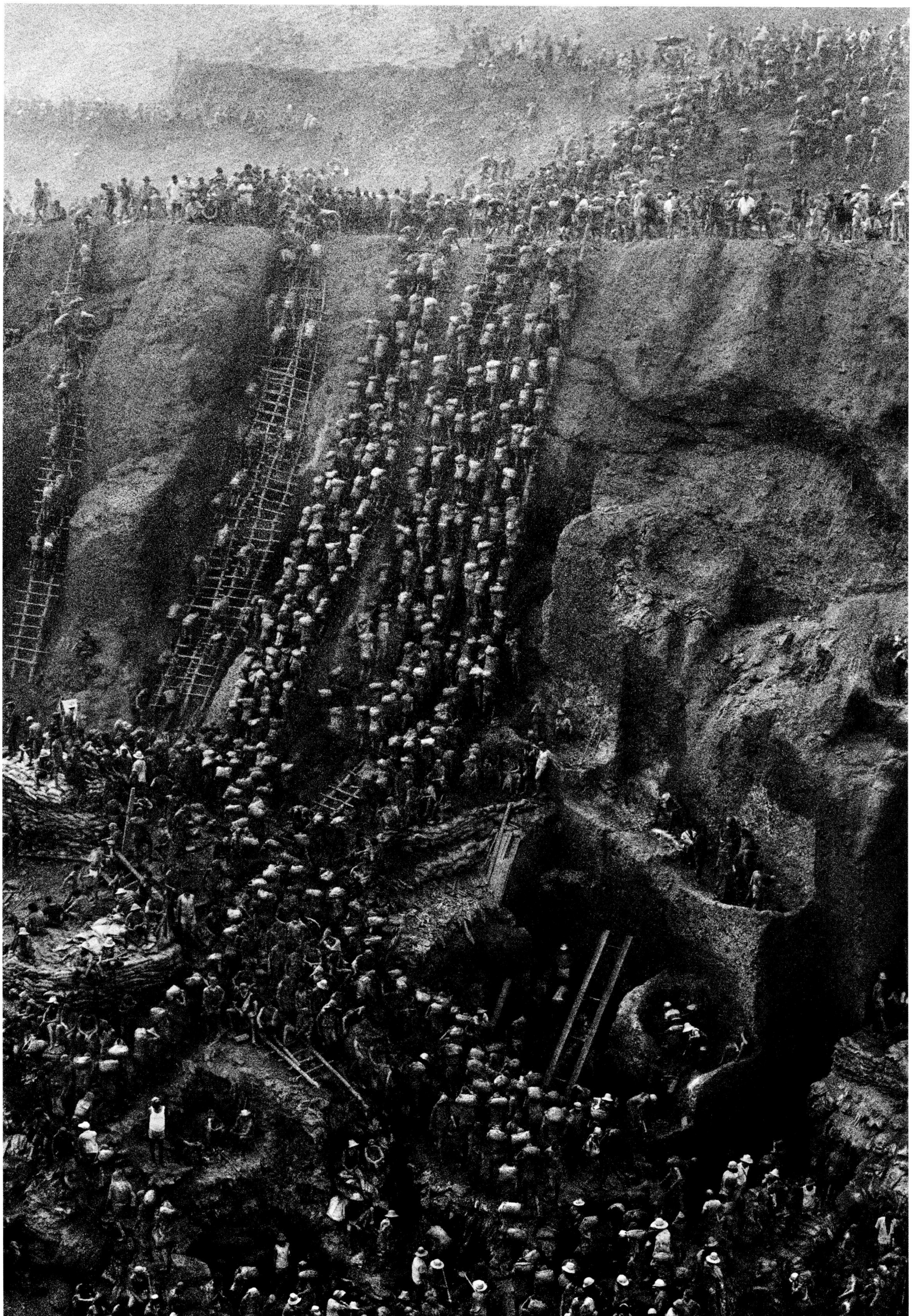

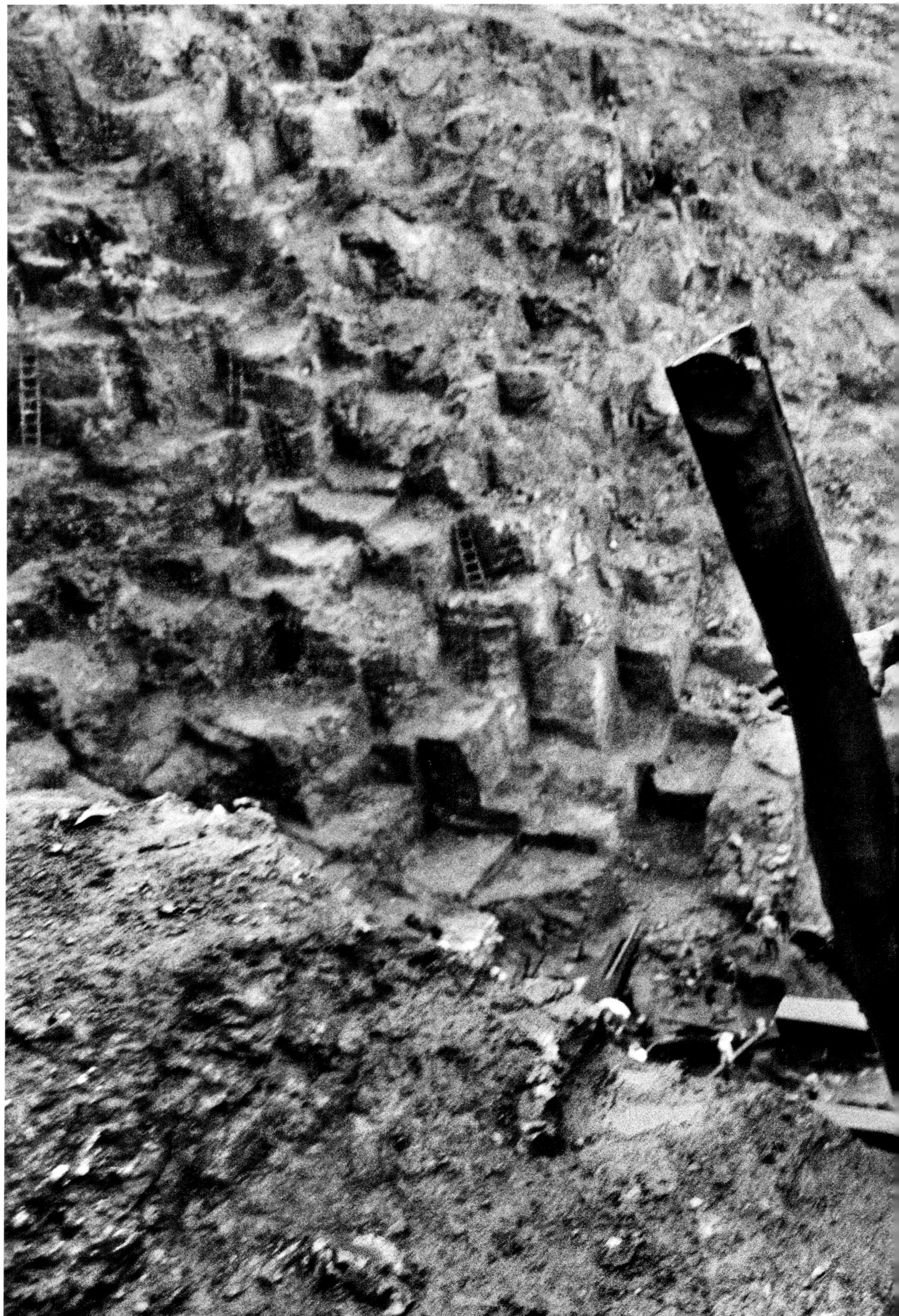

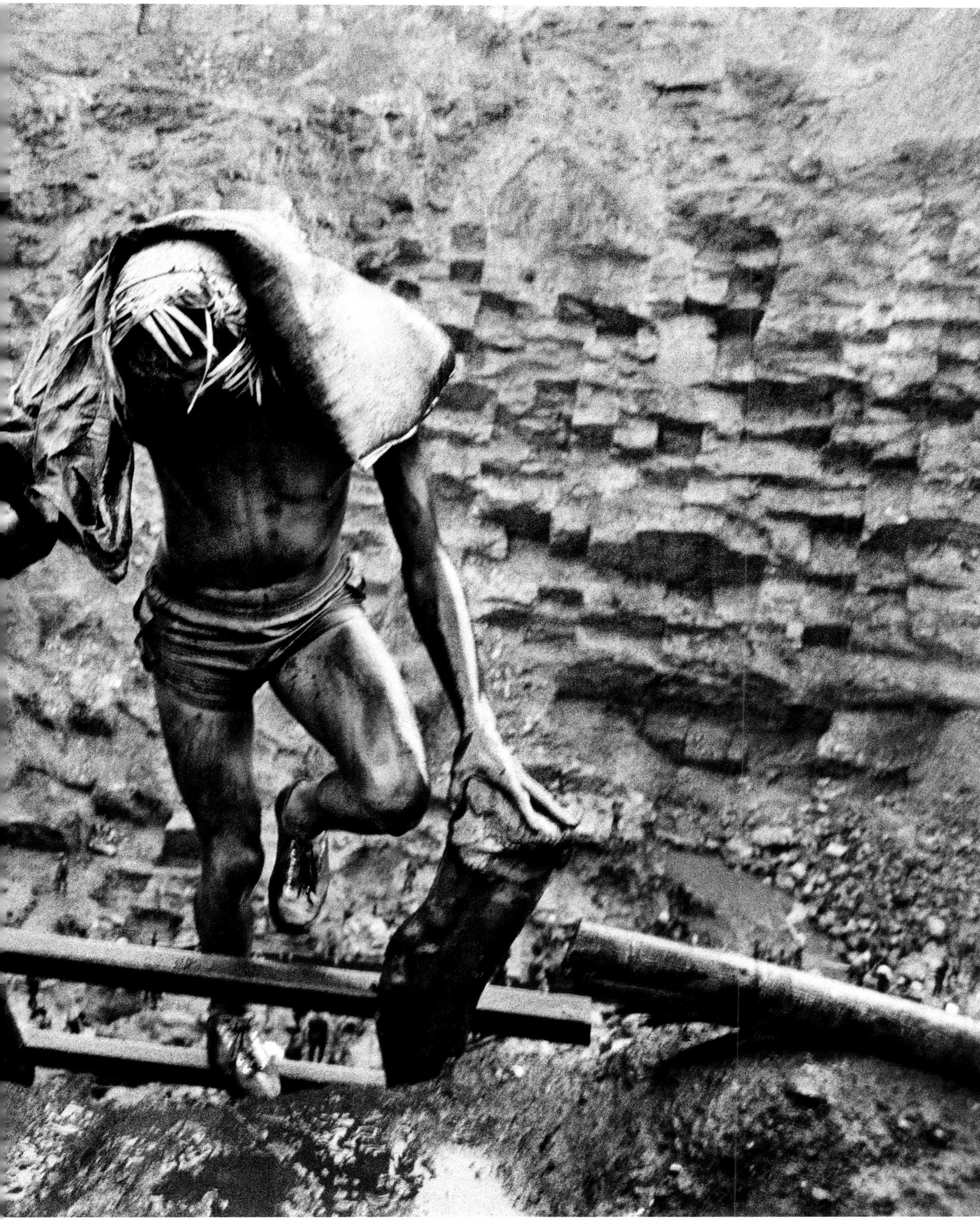

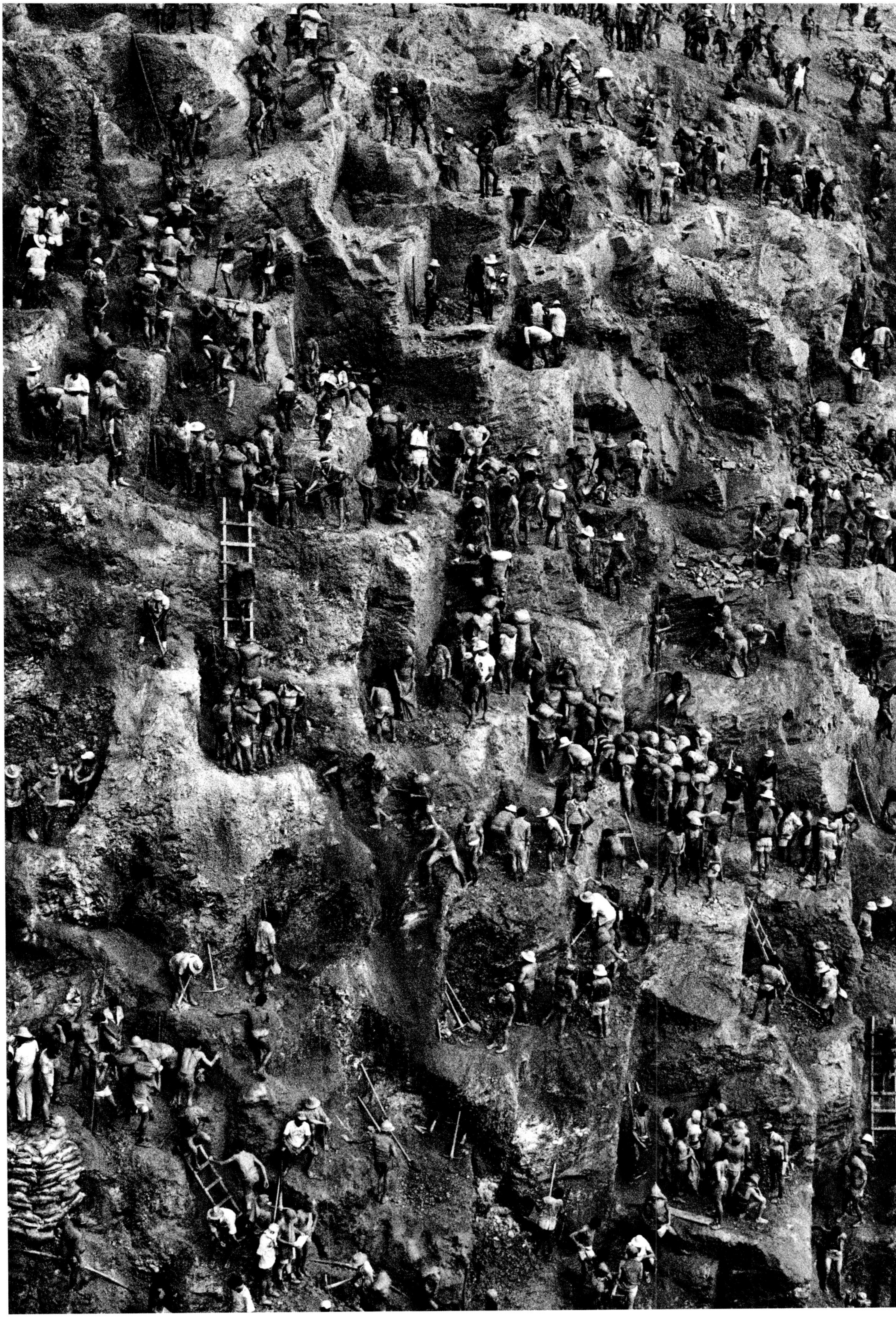

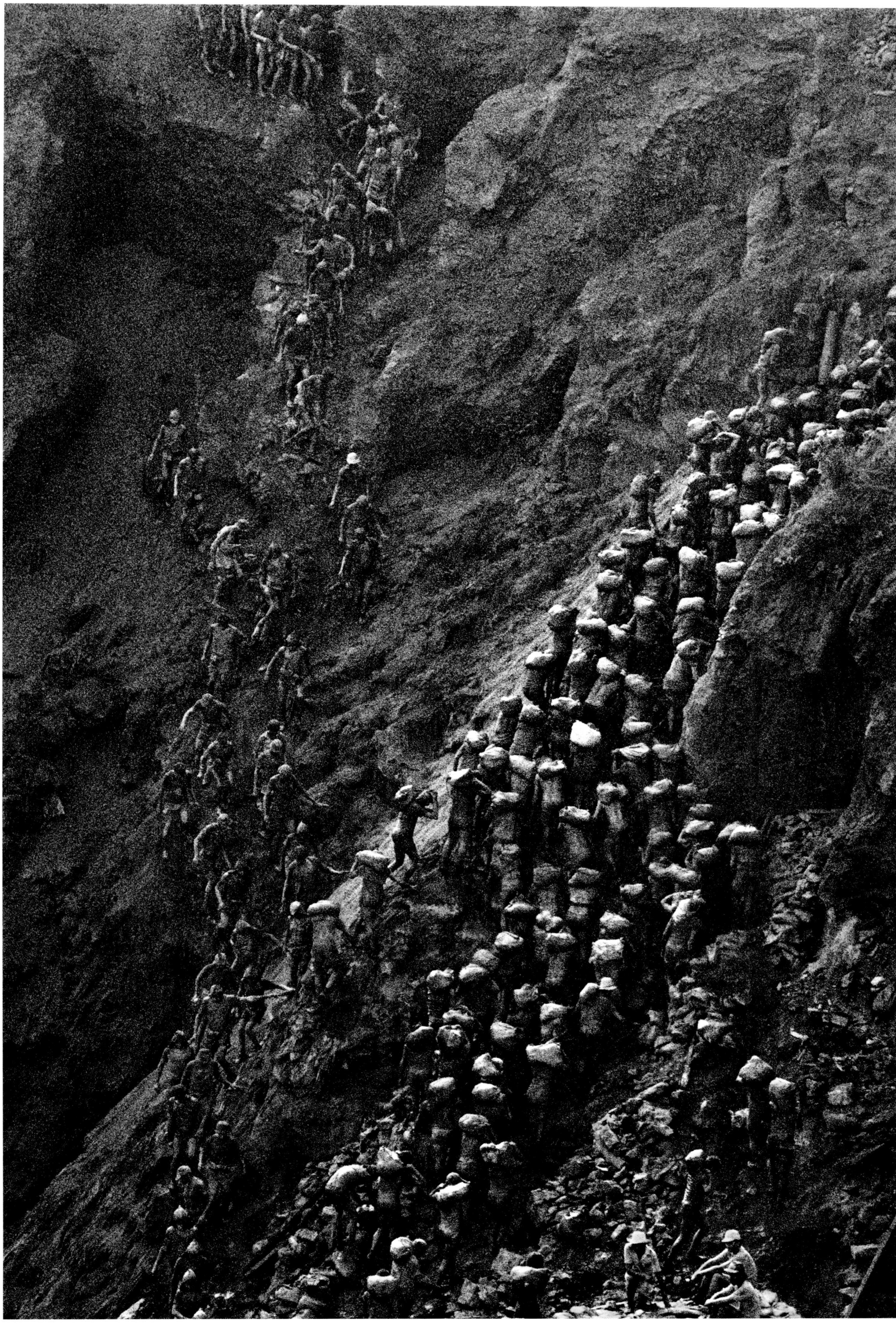

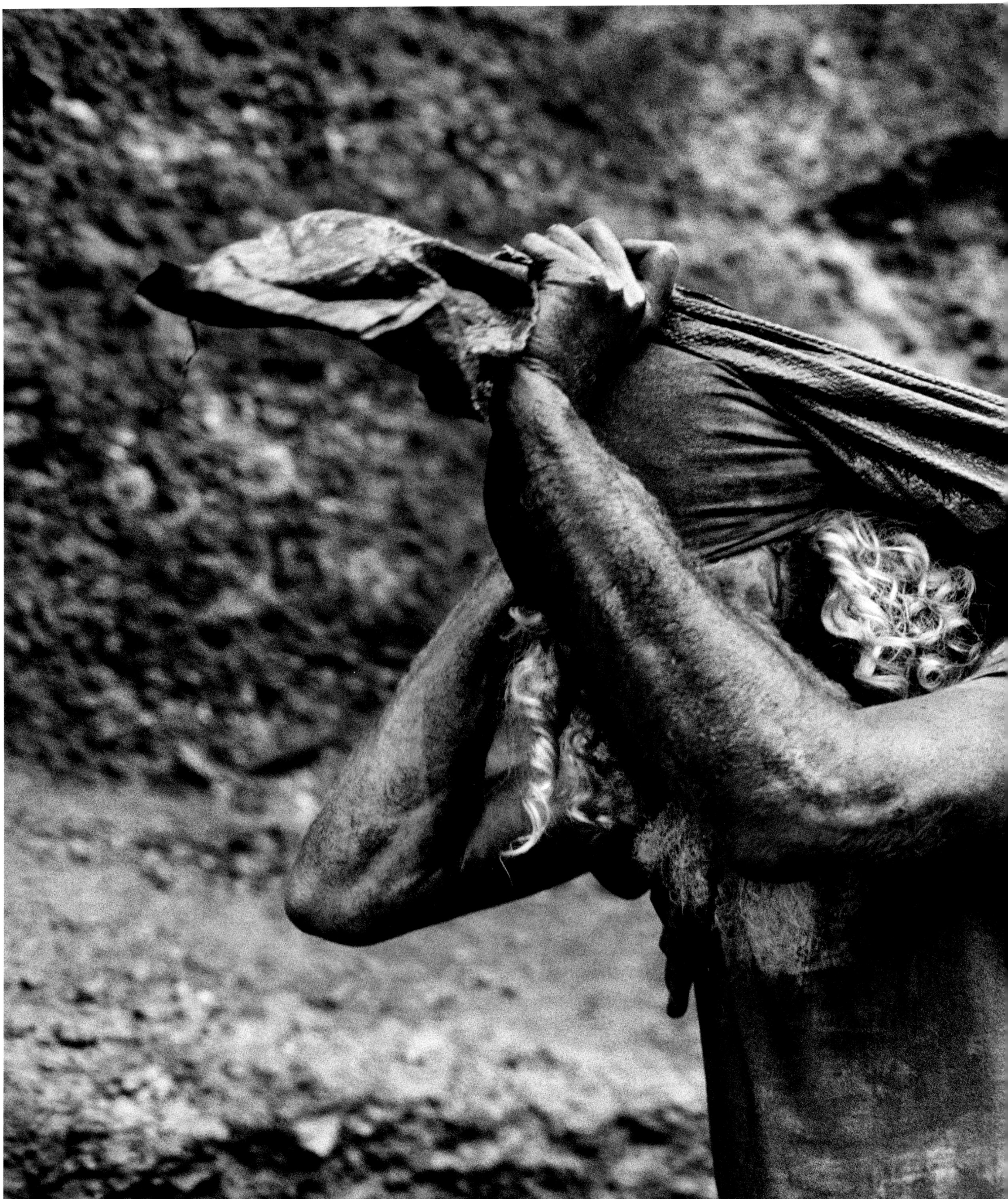

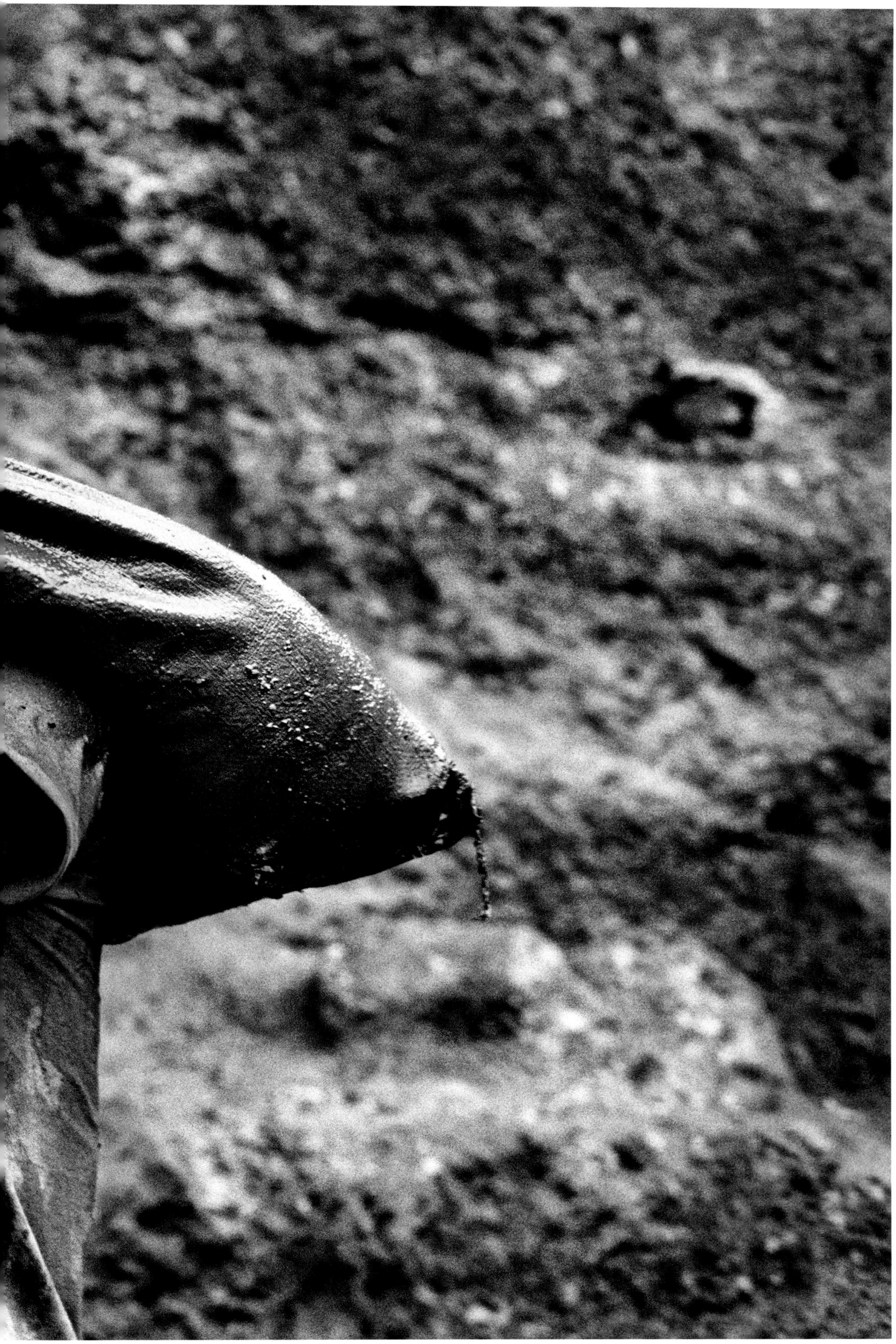

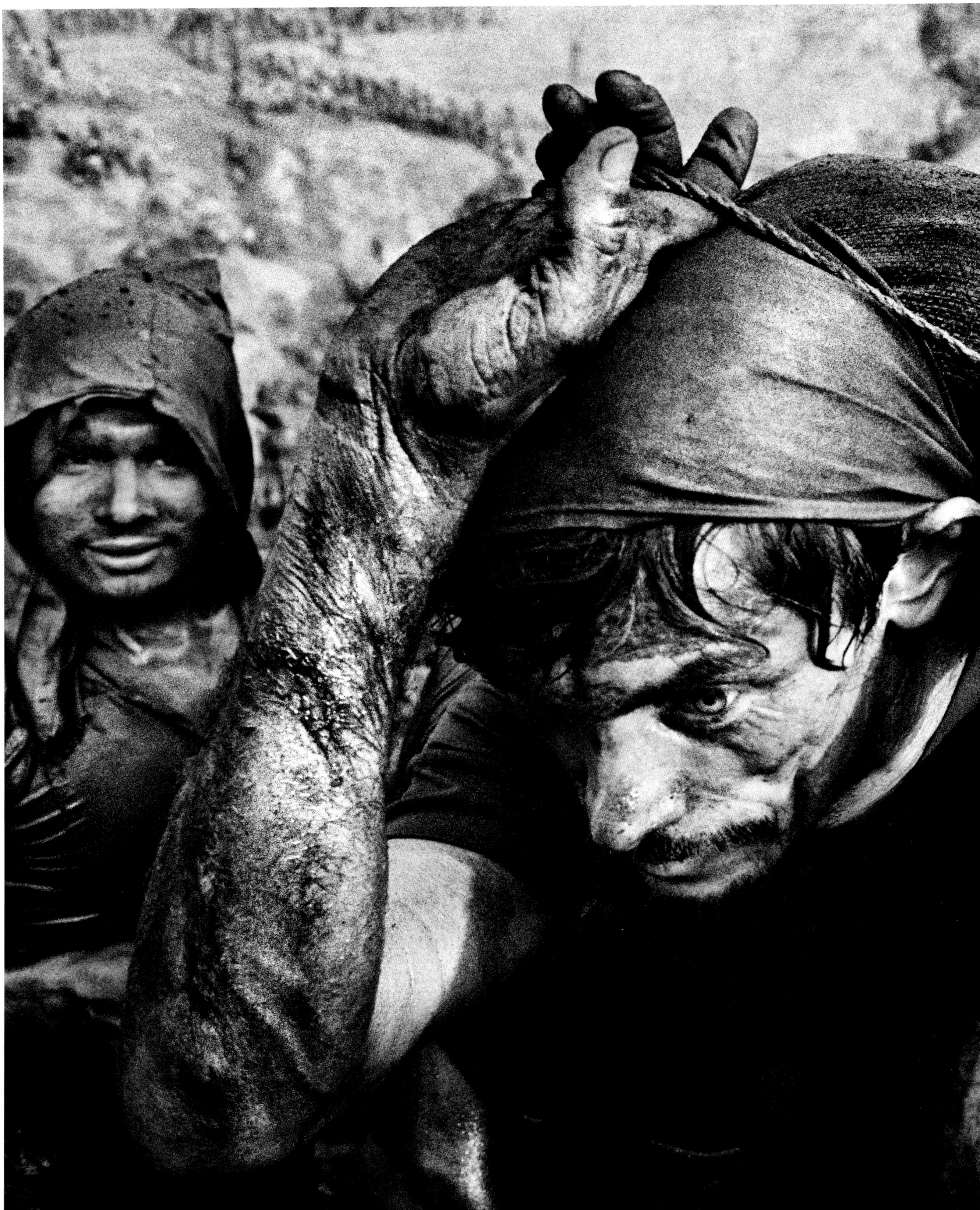

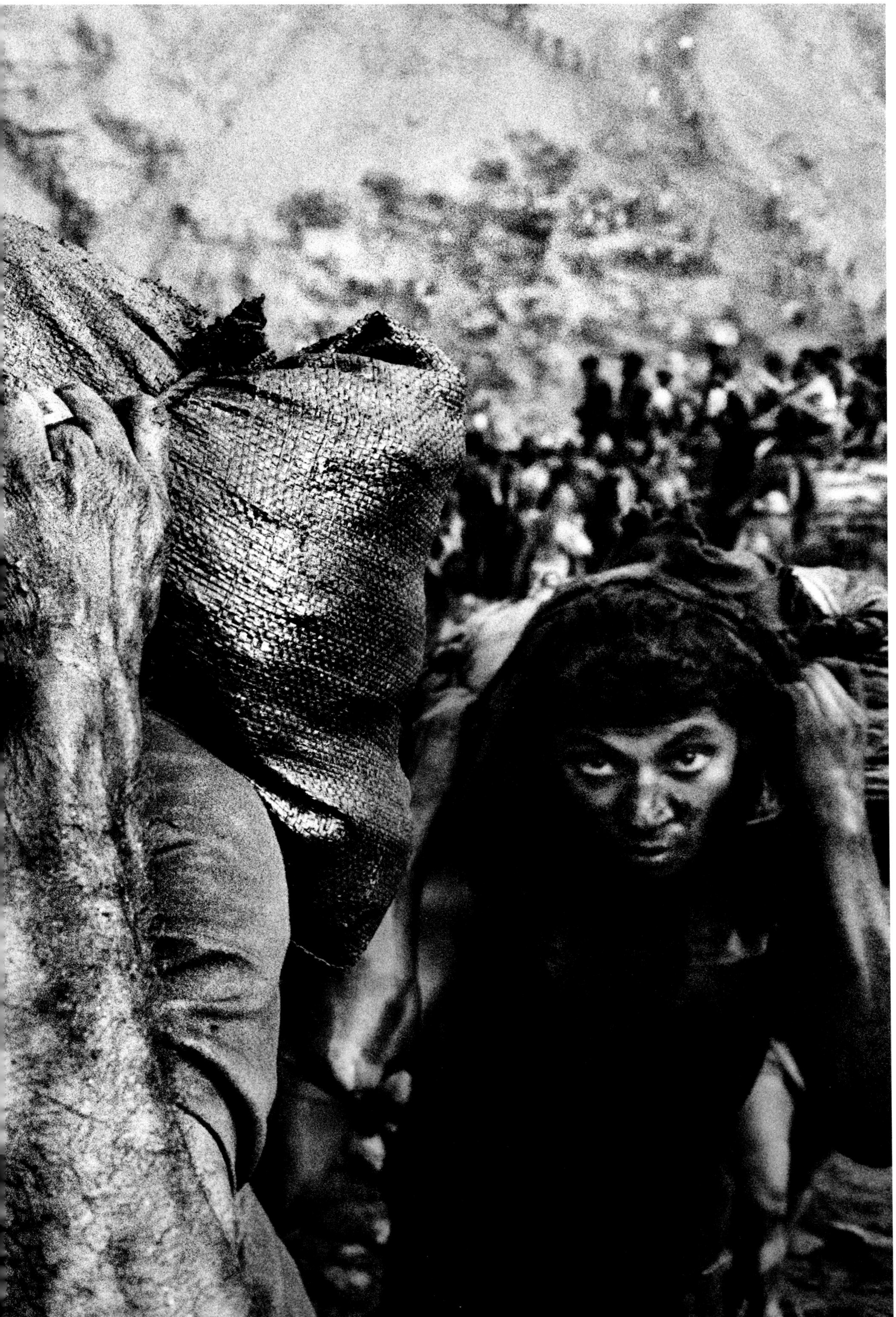

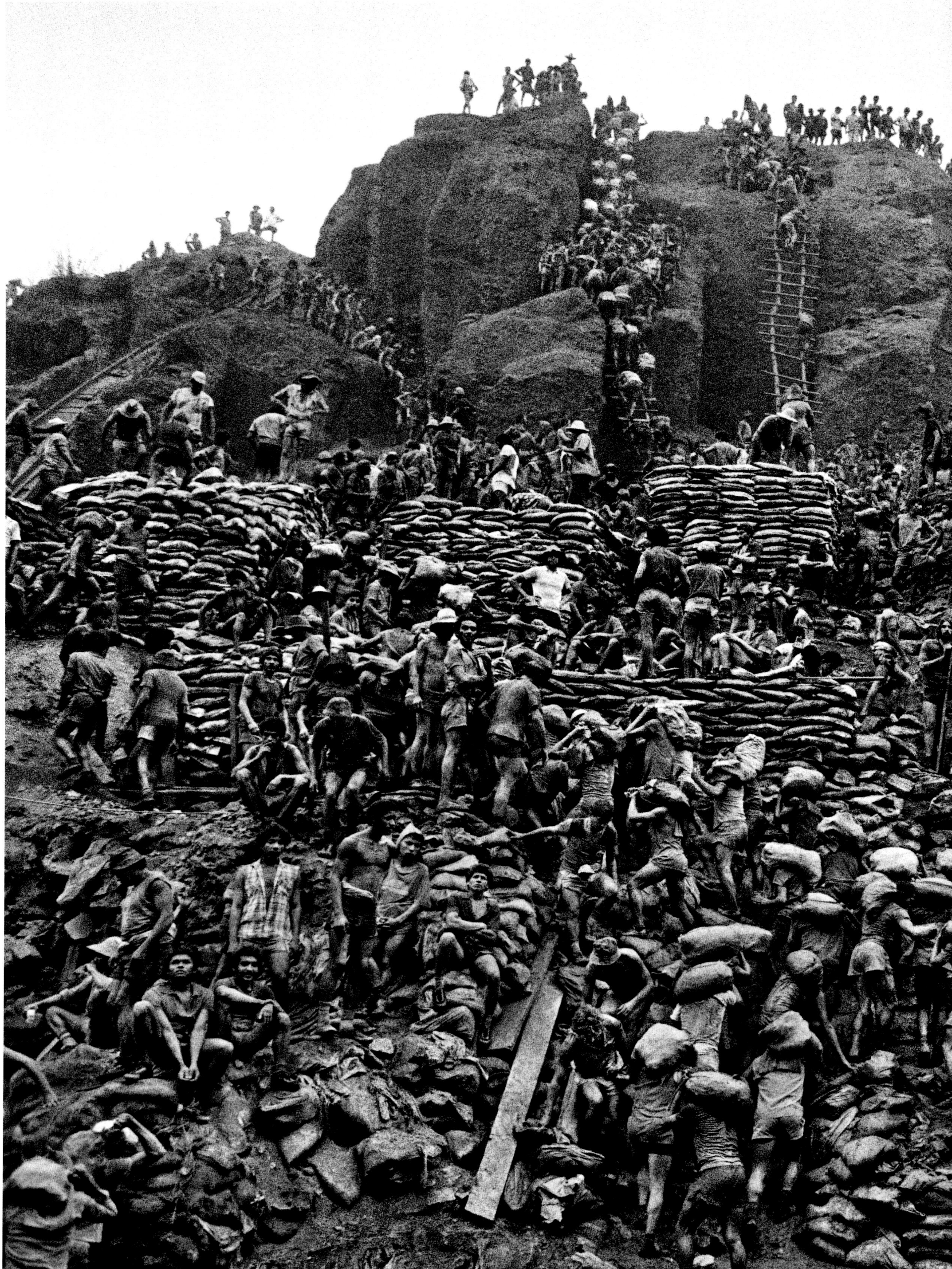

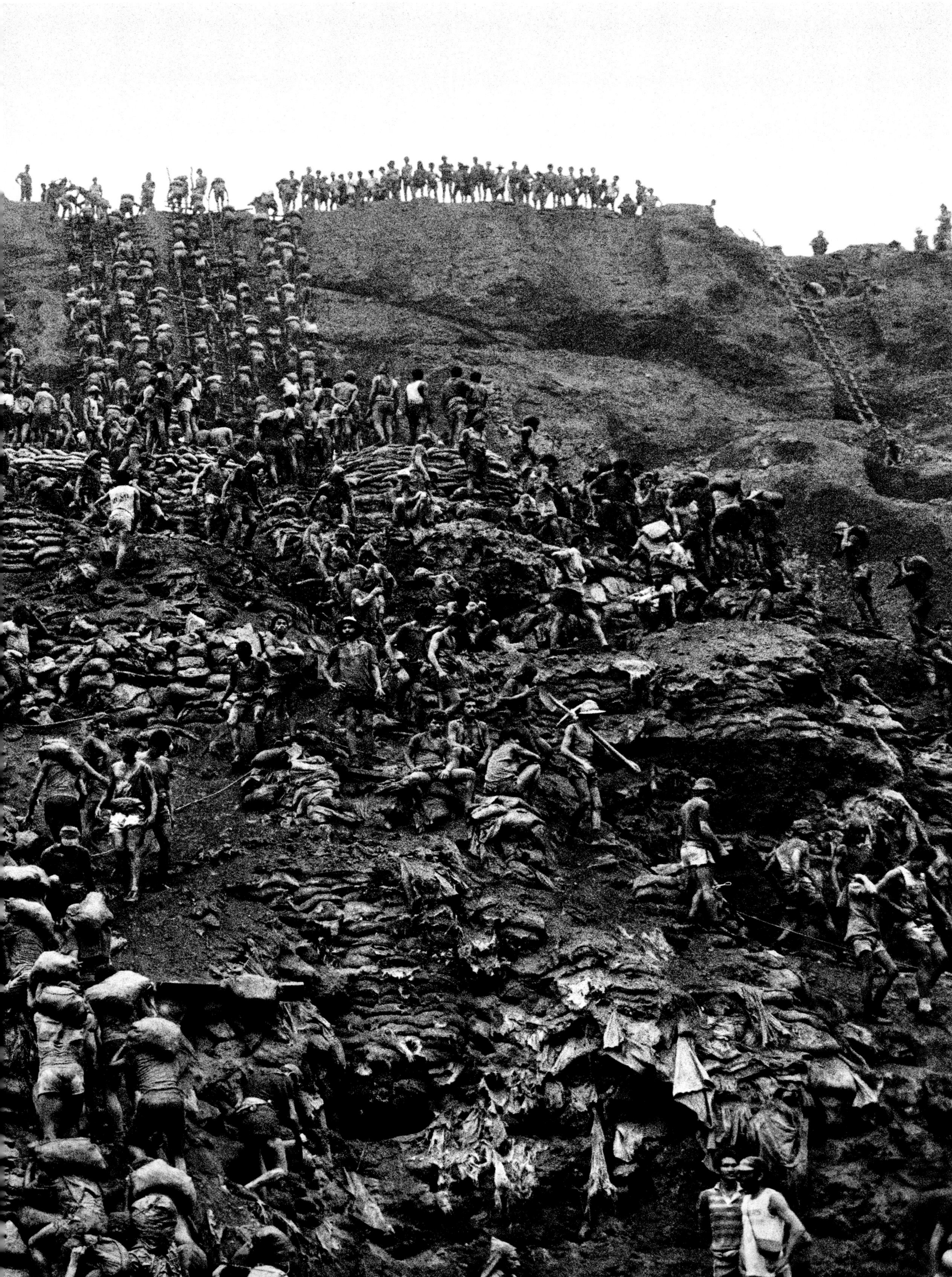

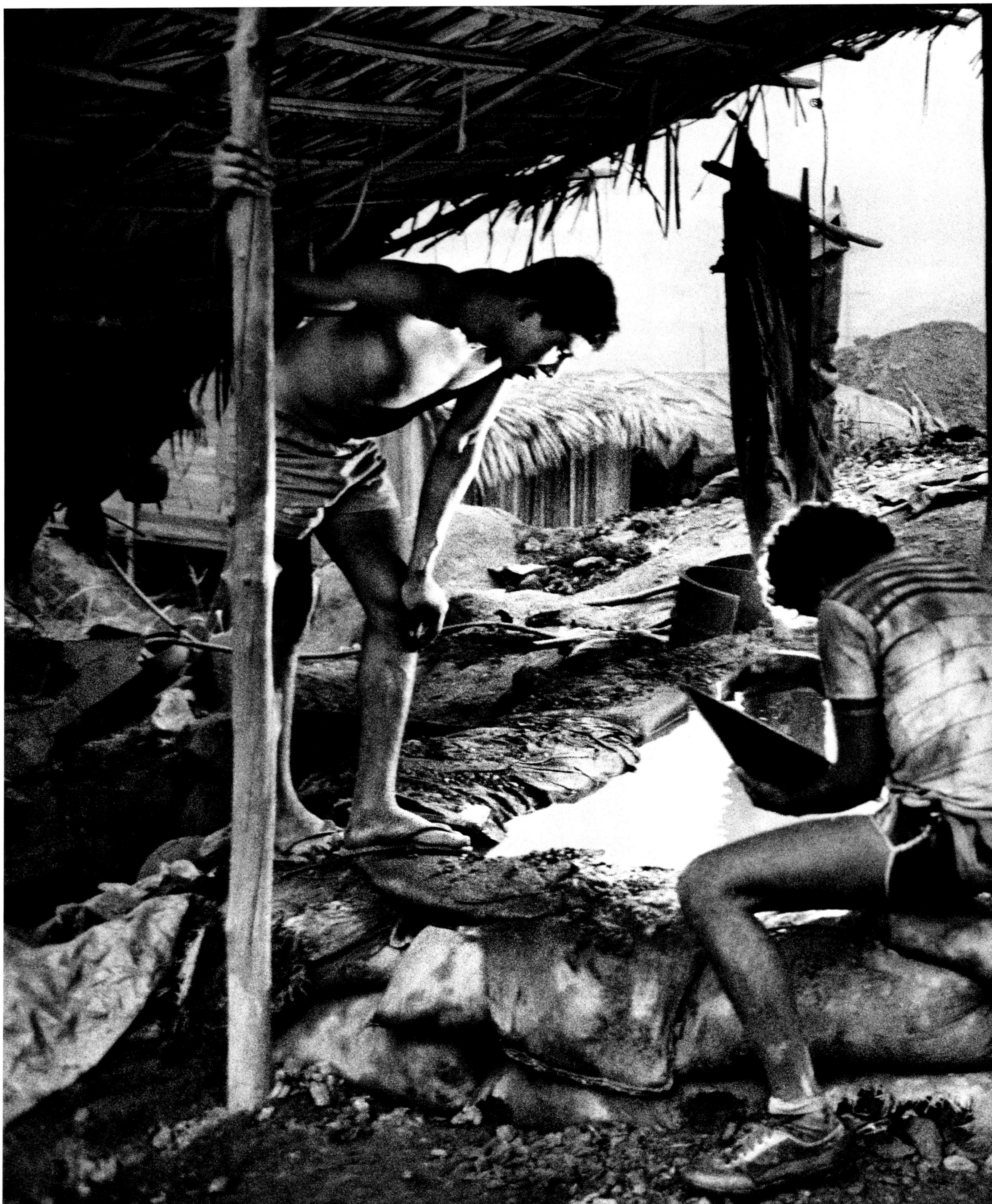

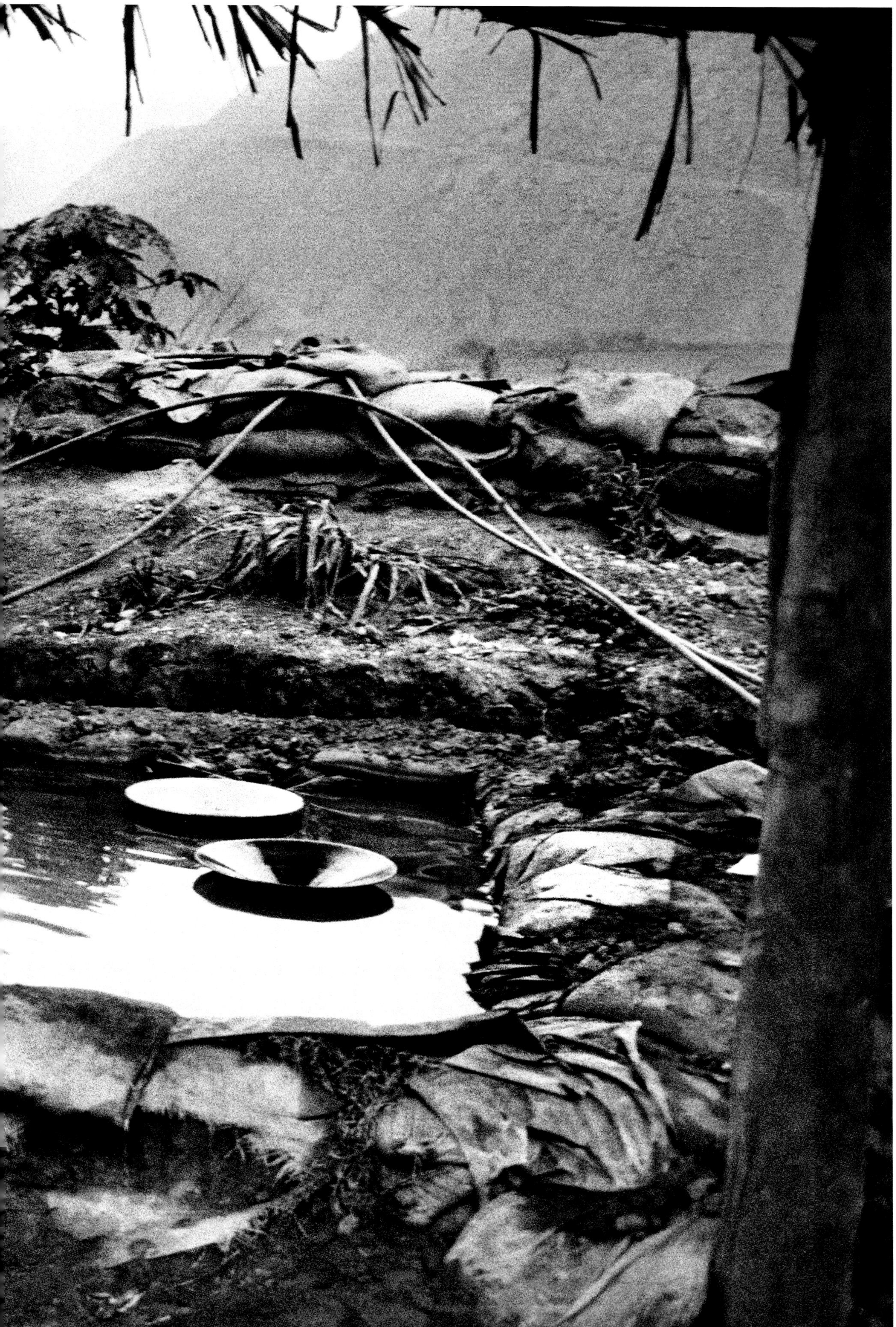

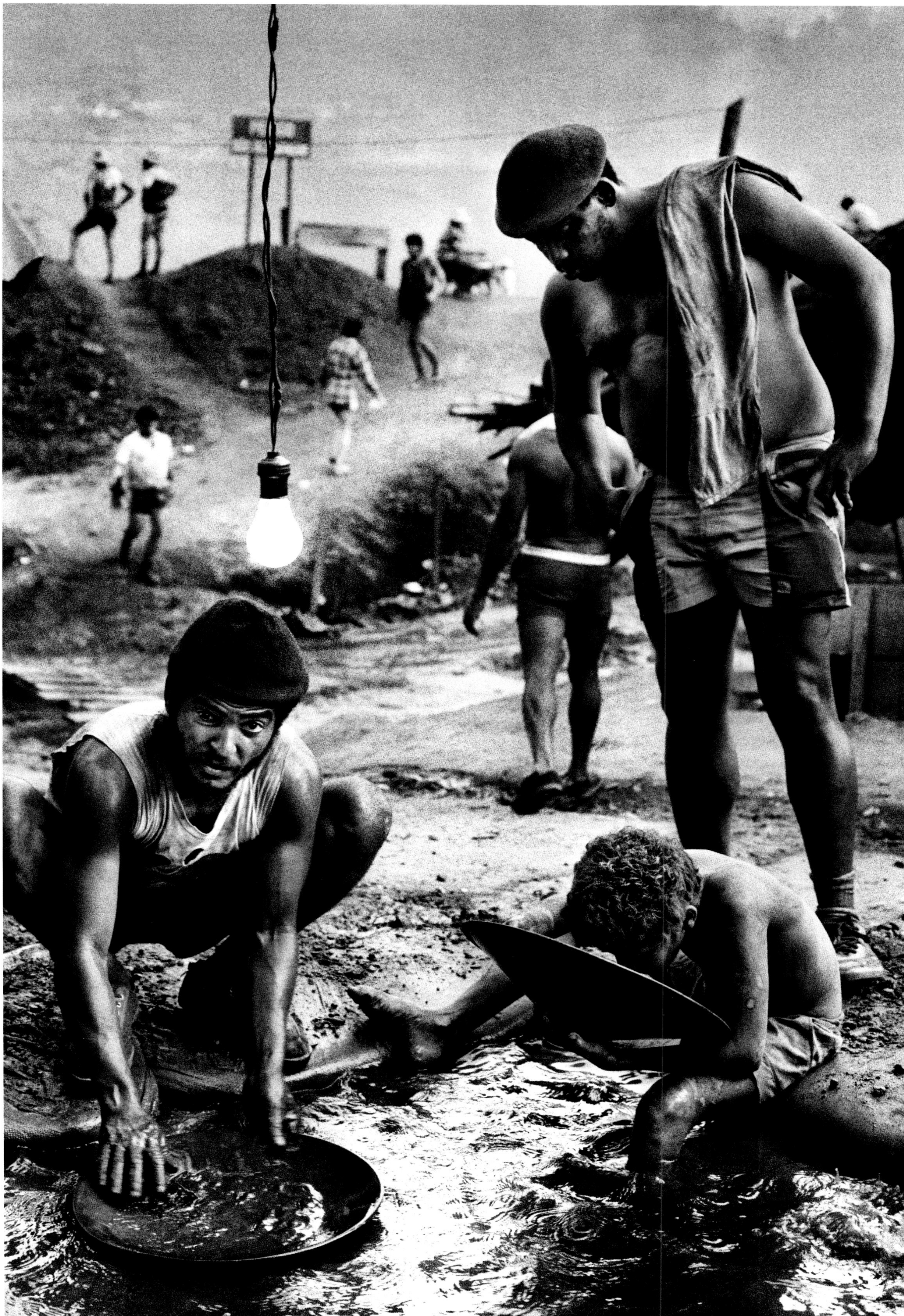

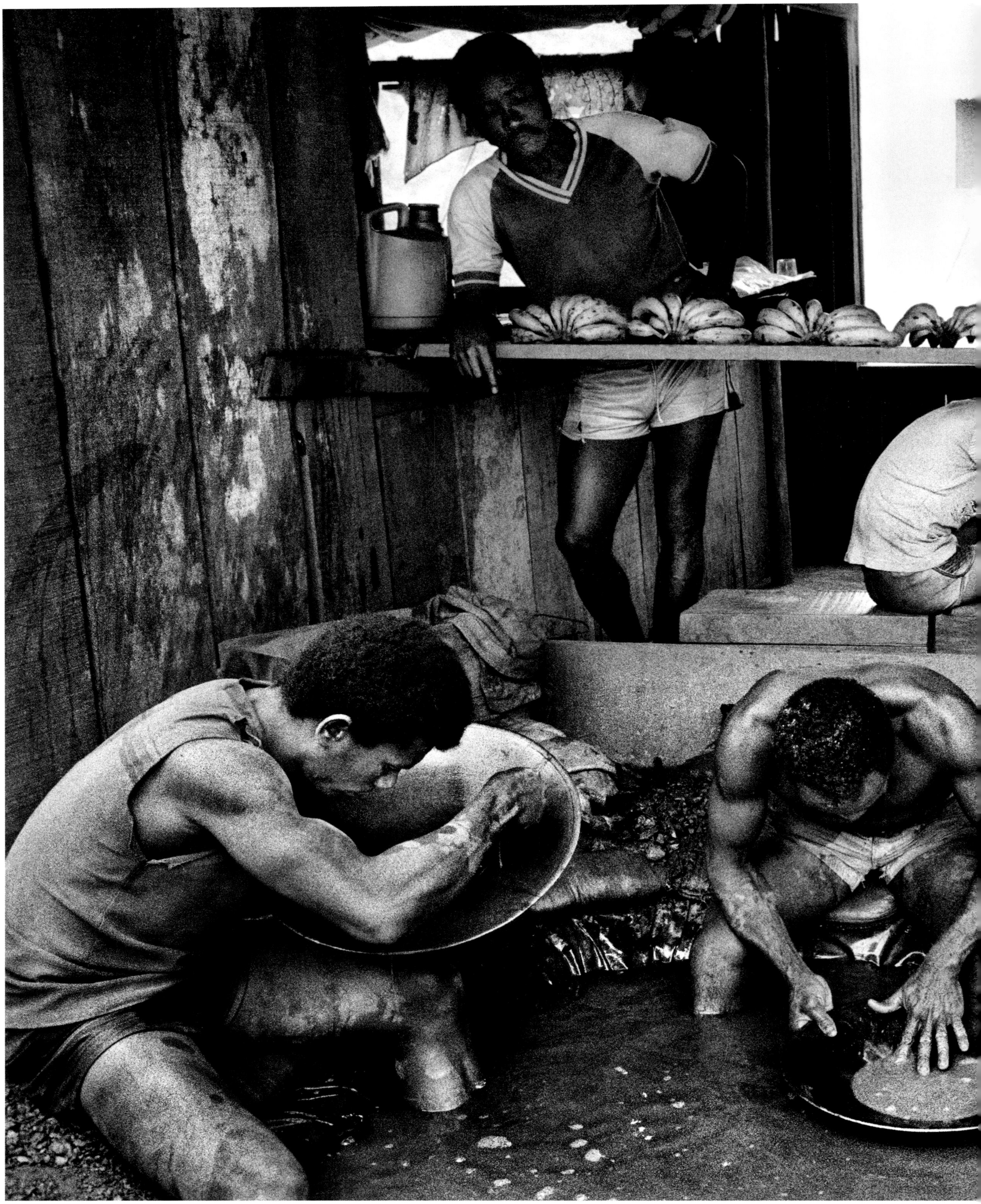

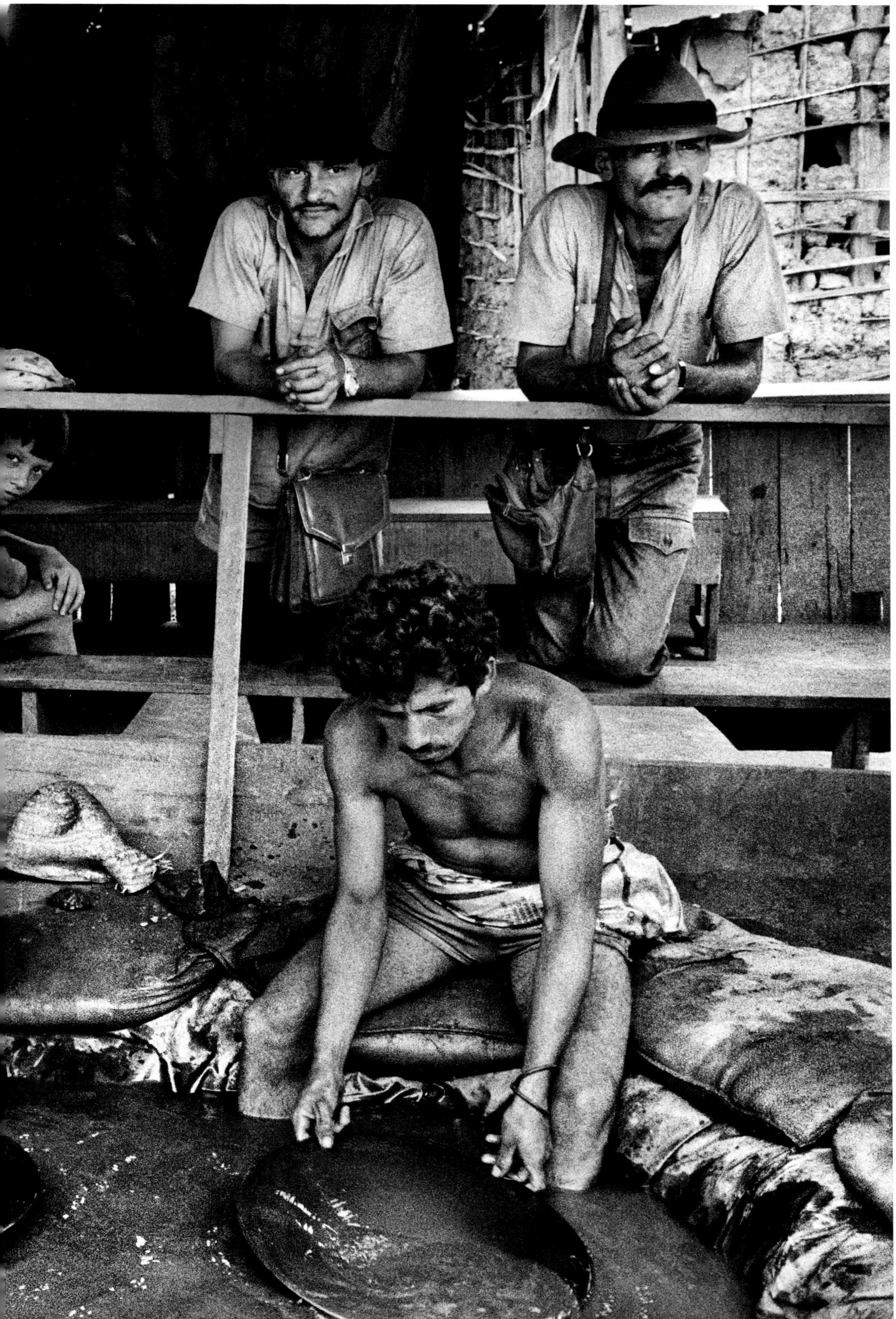

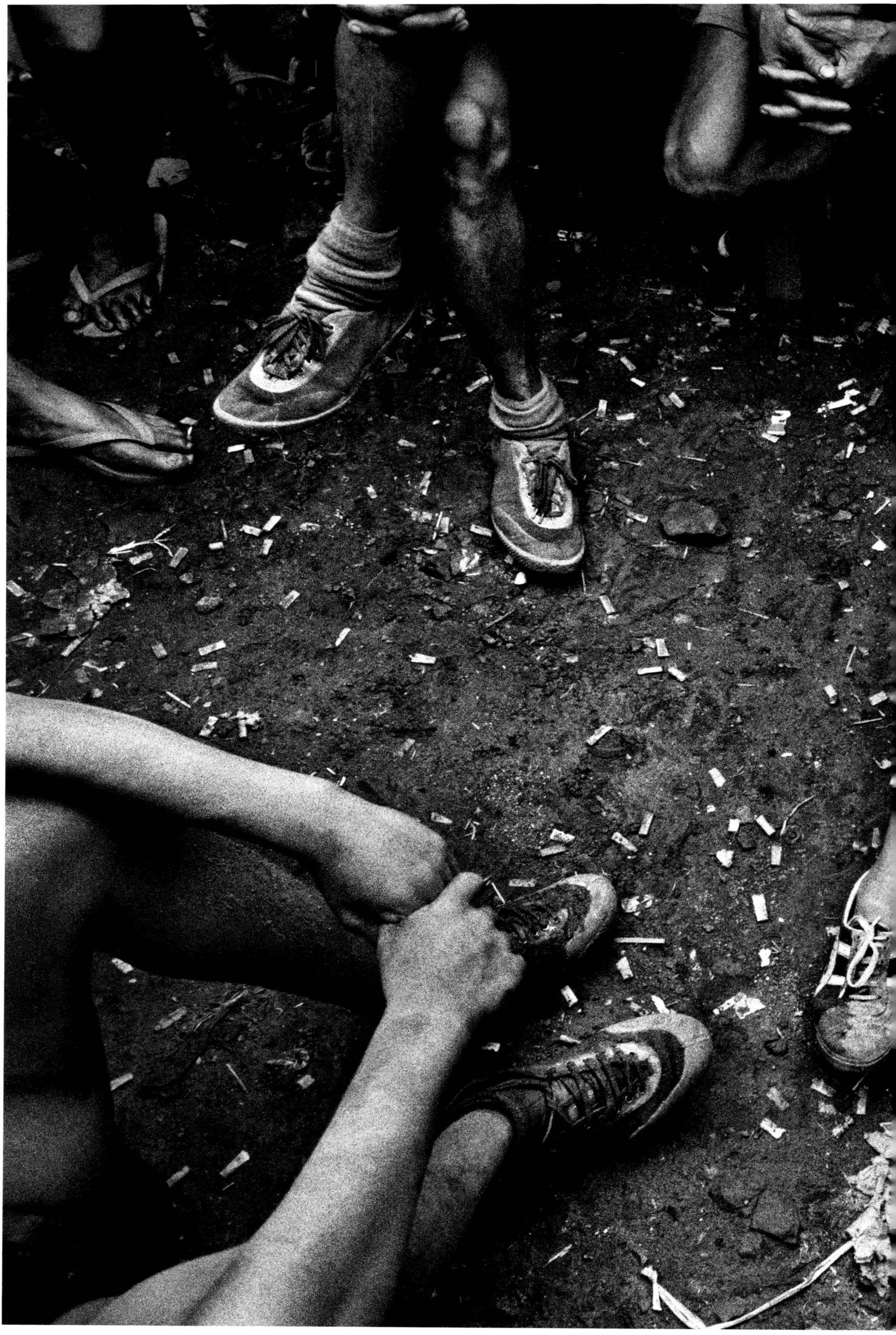

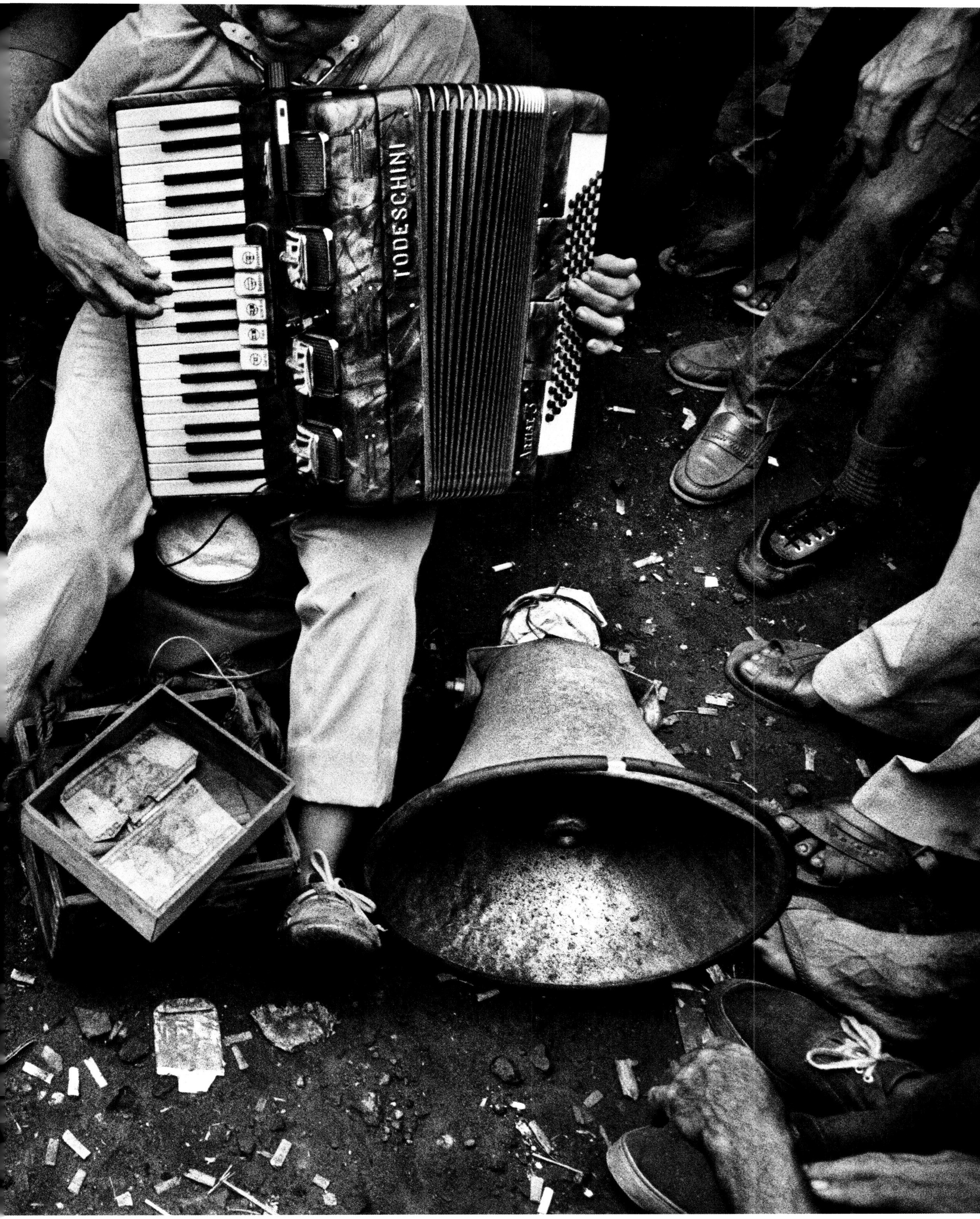
TODESCHINI

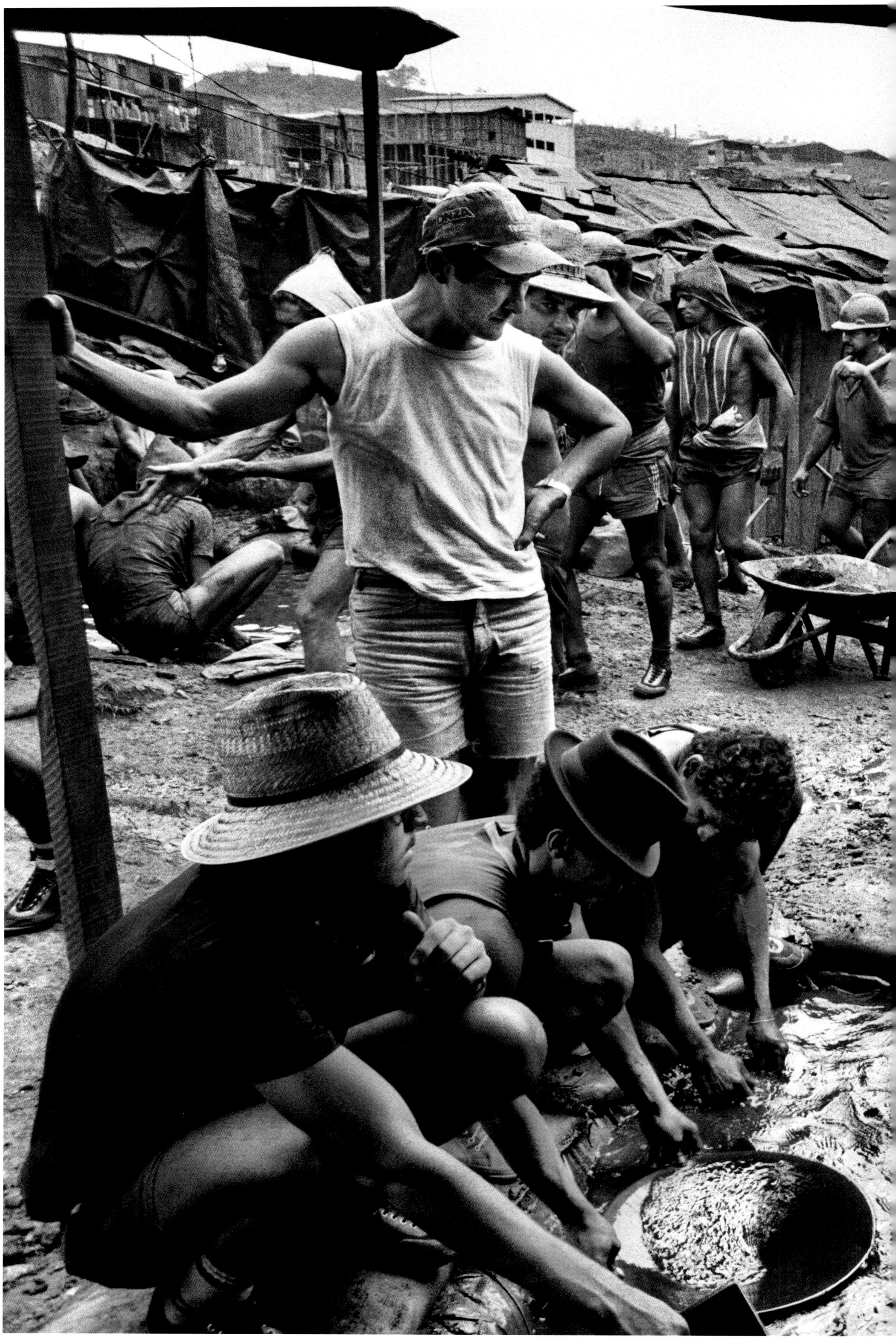

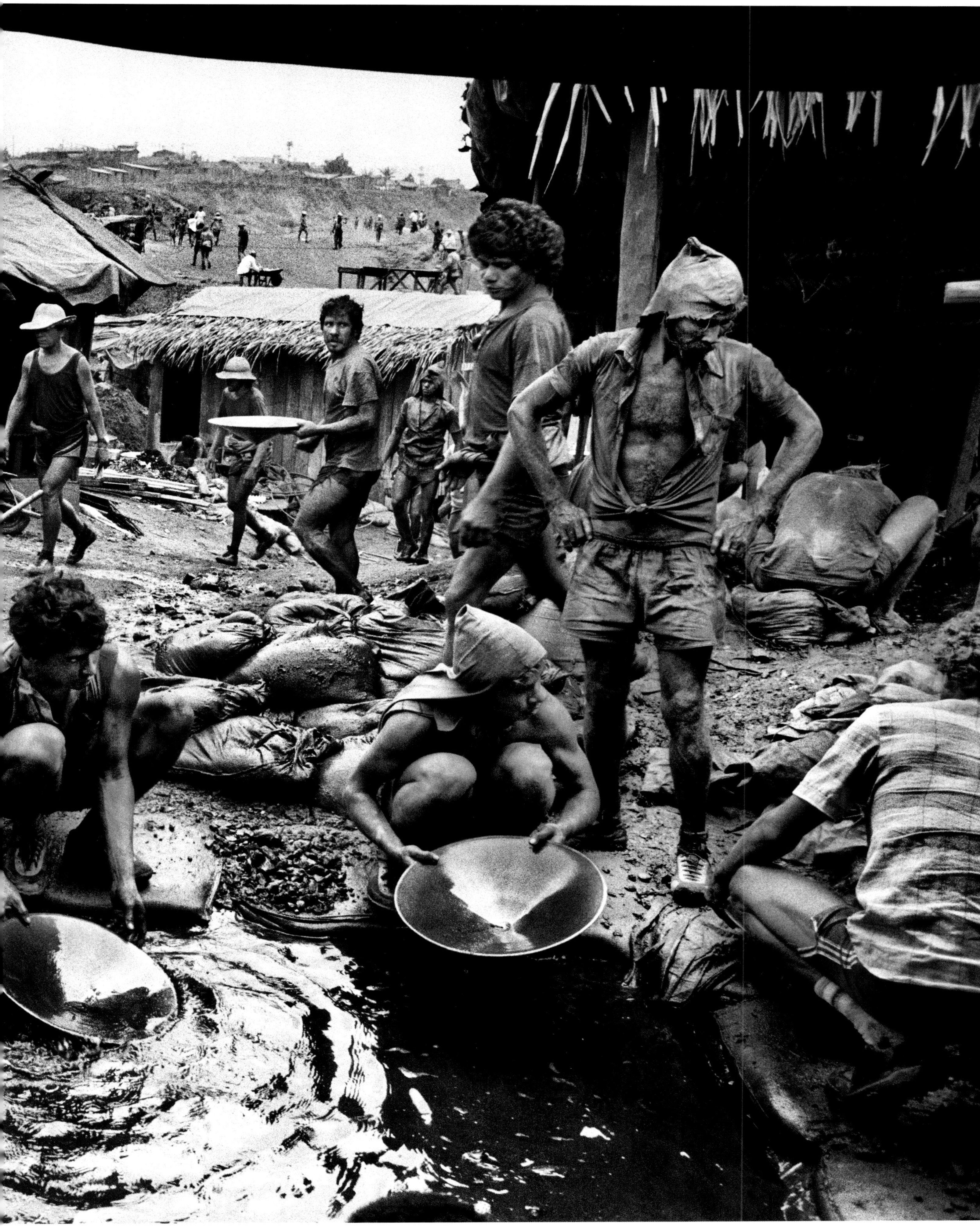

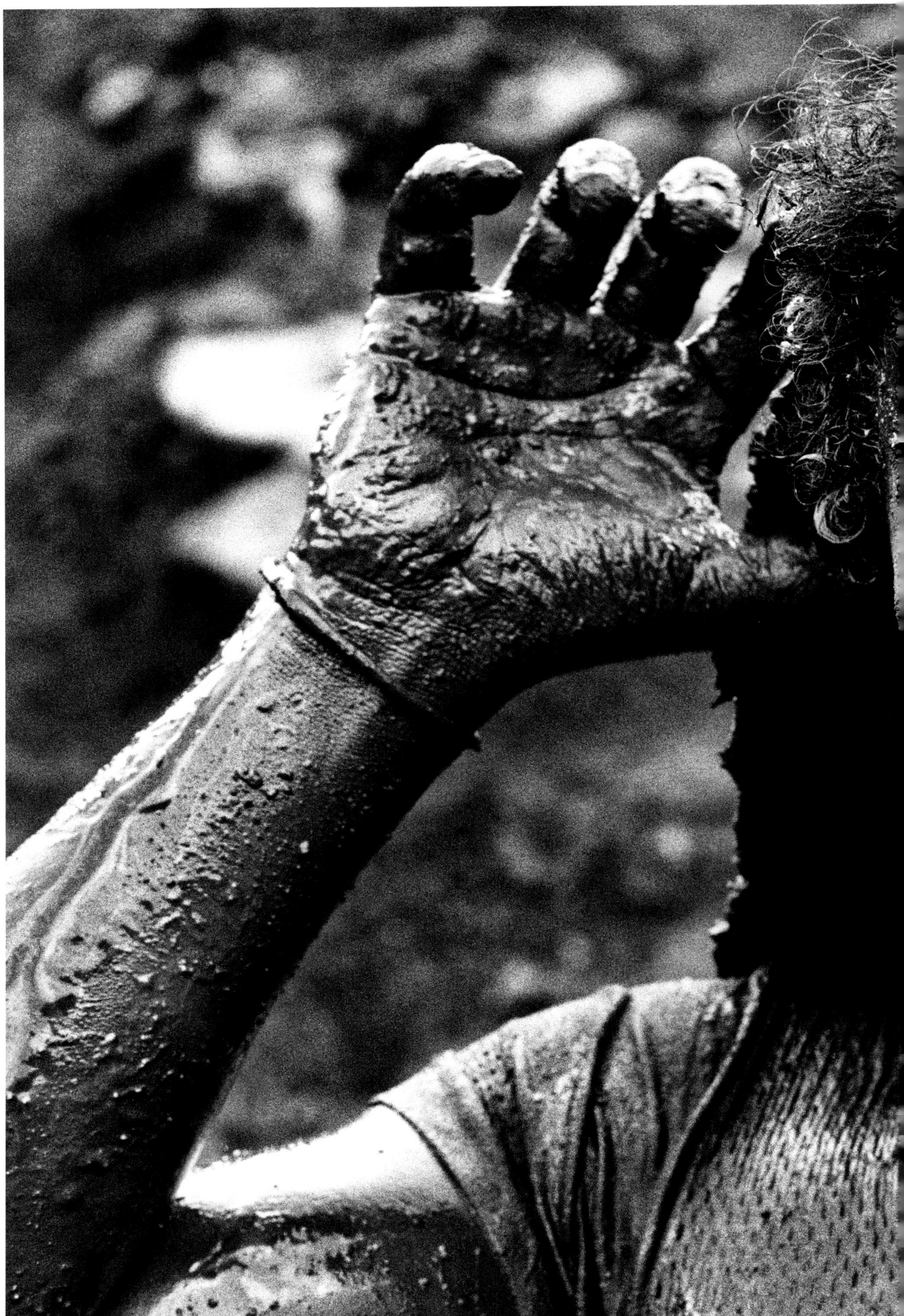

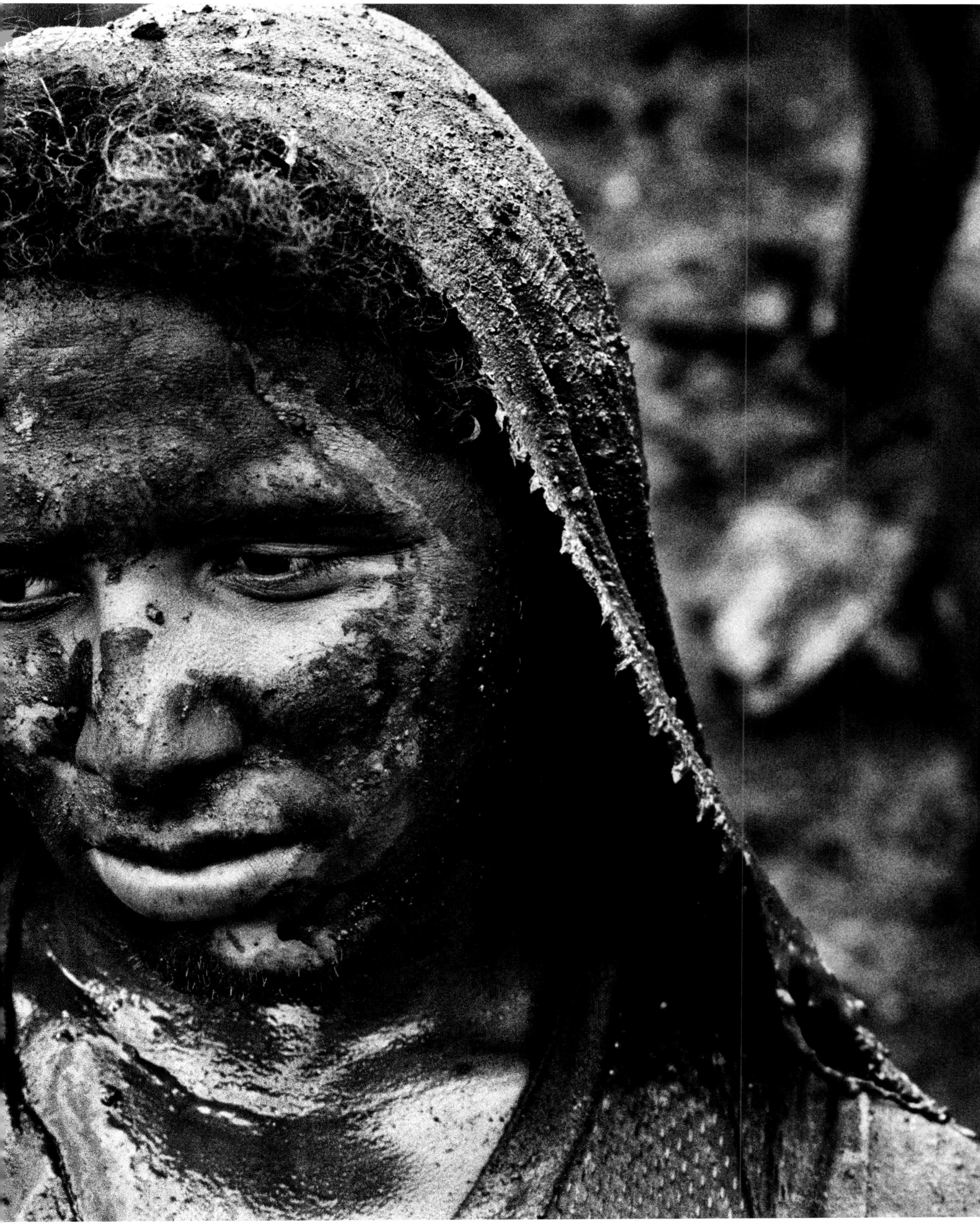

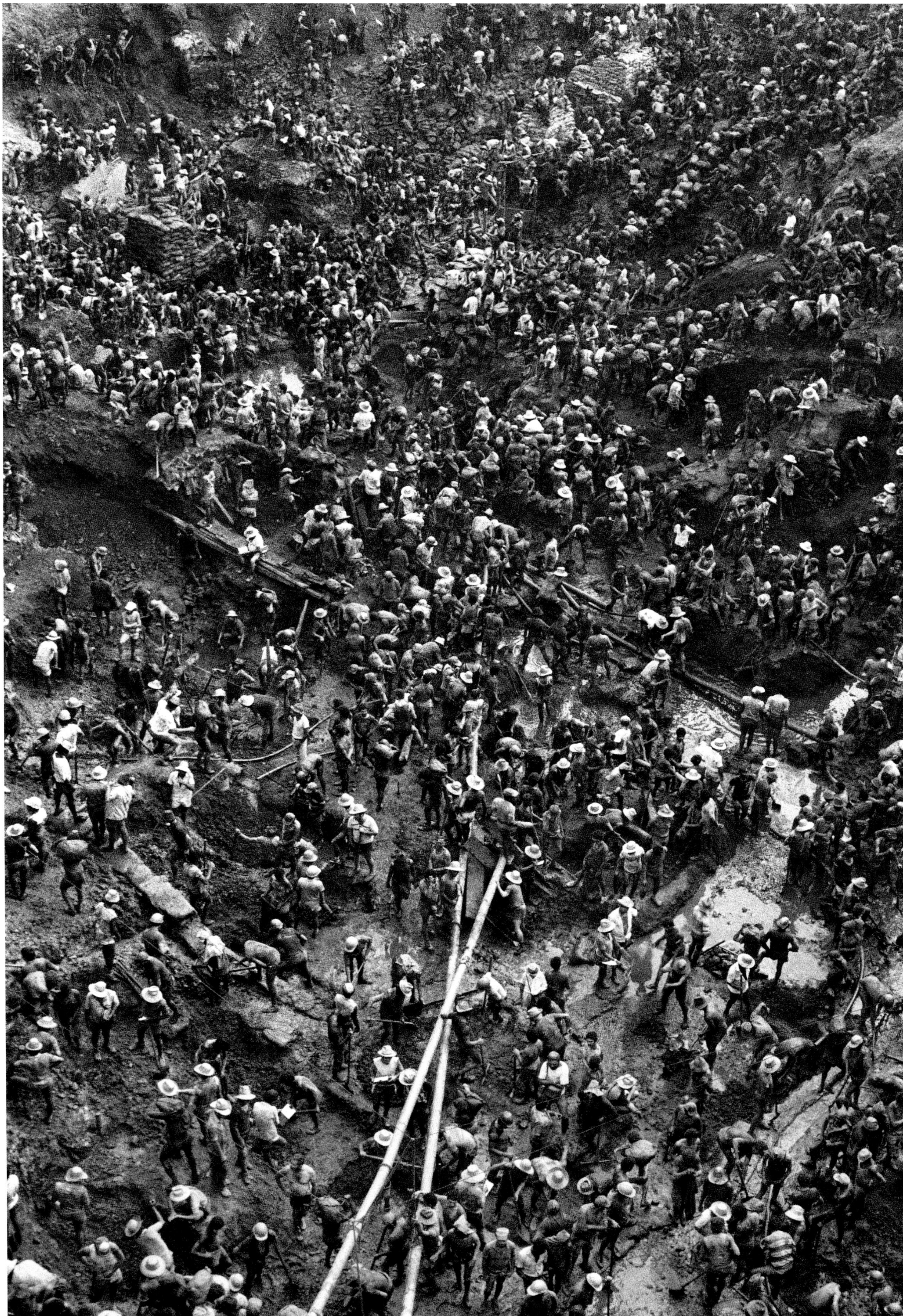

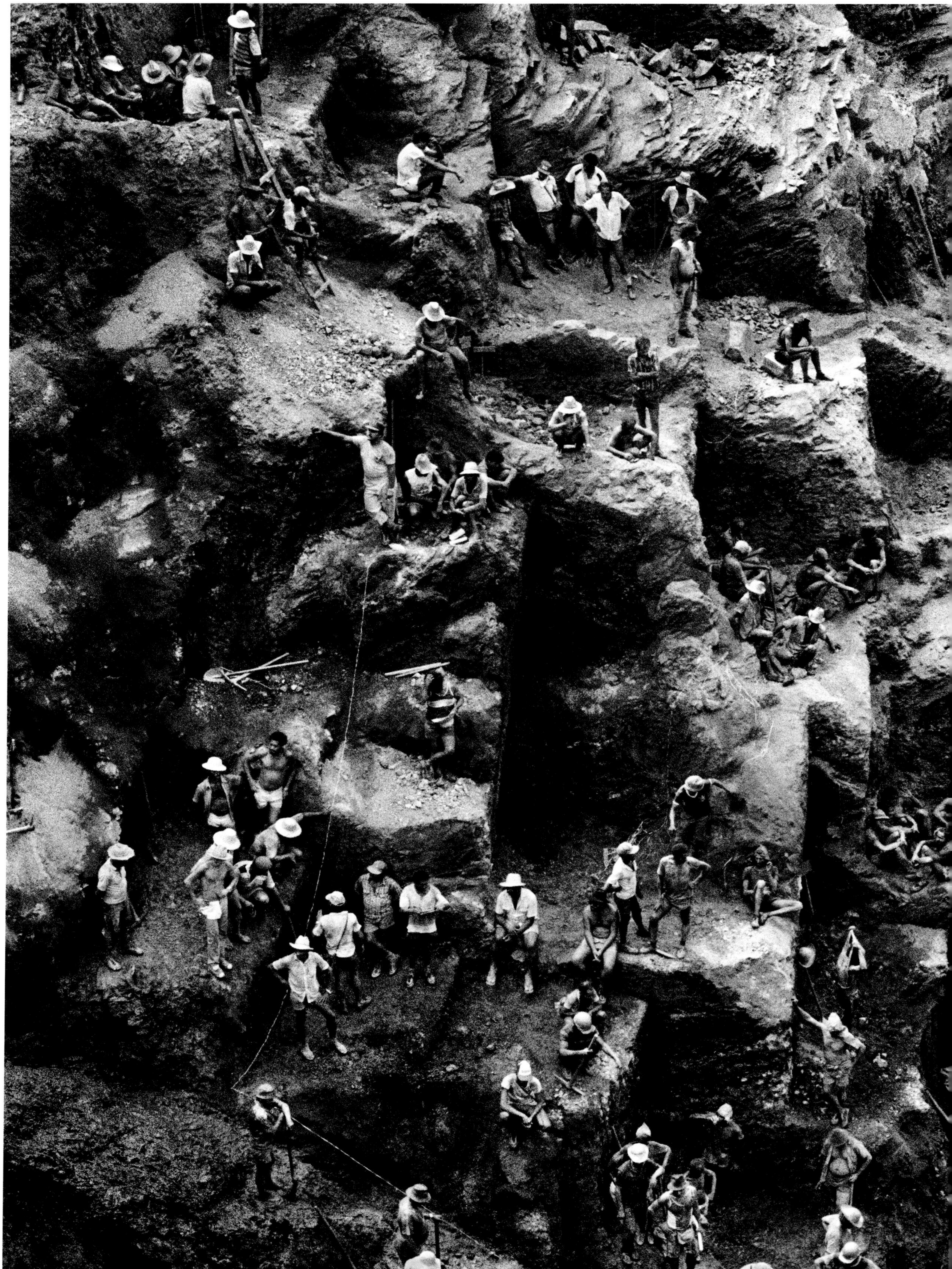

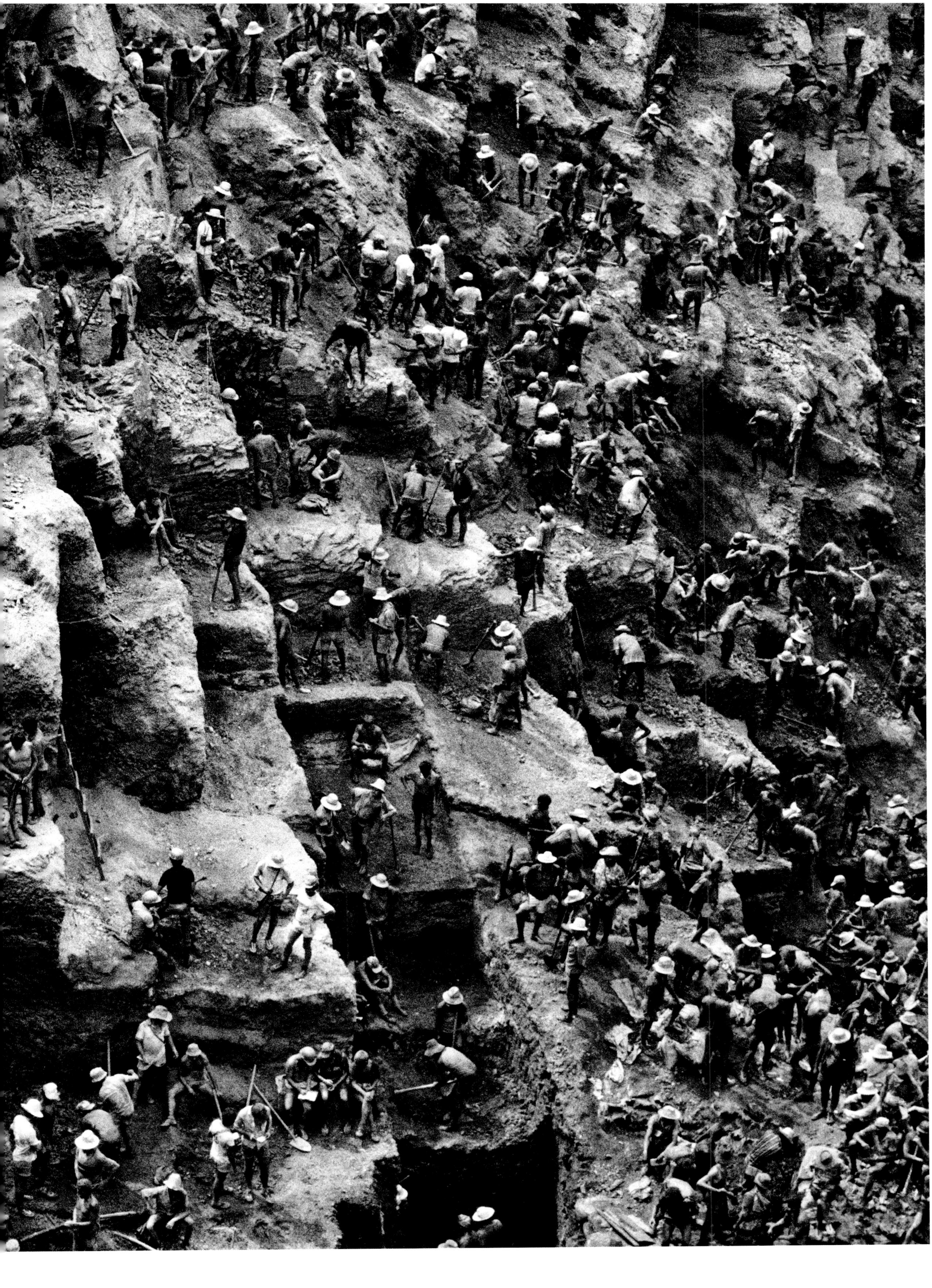

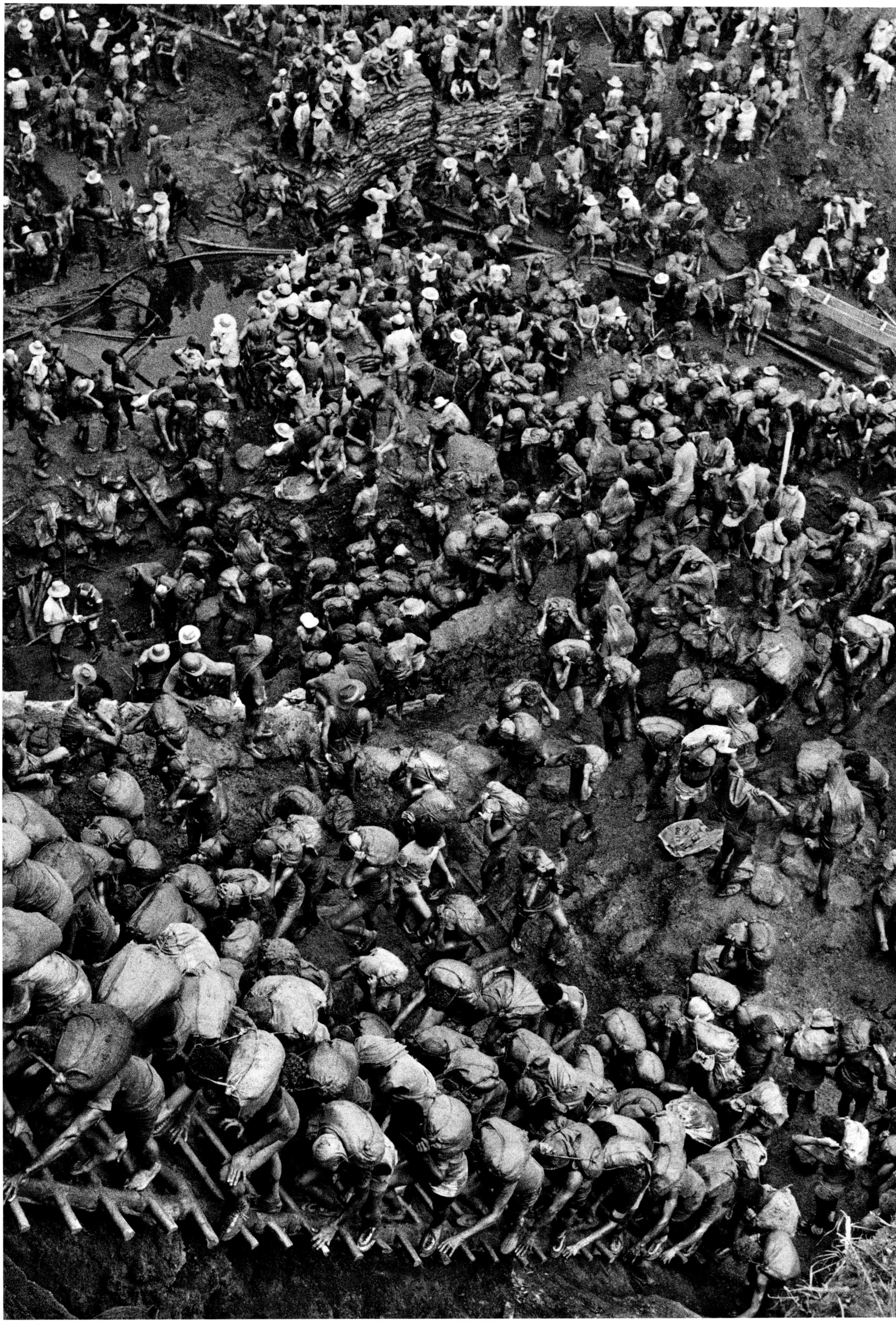

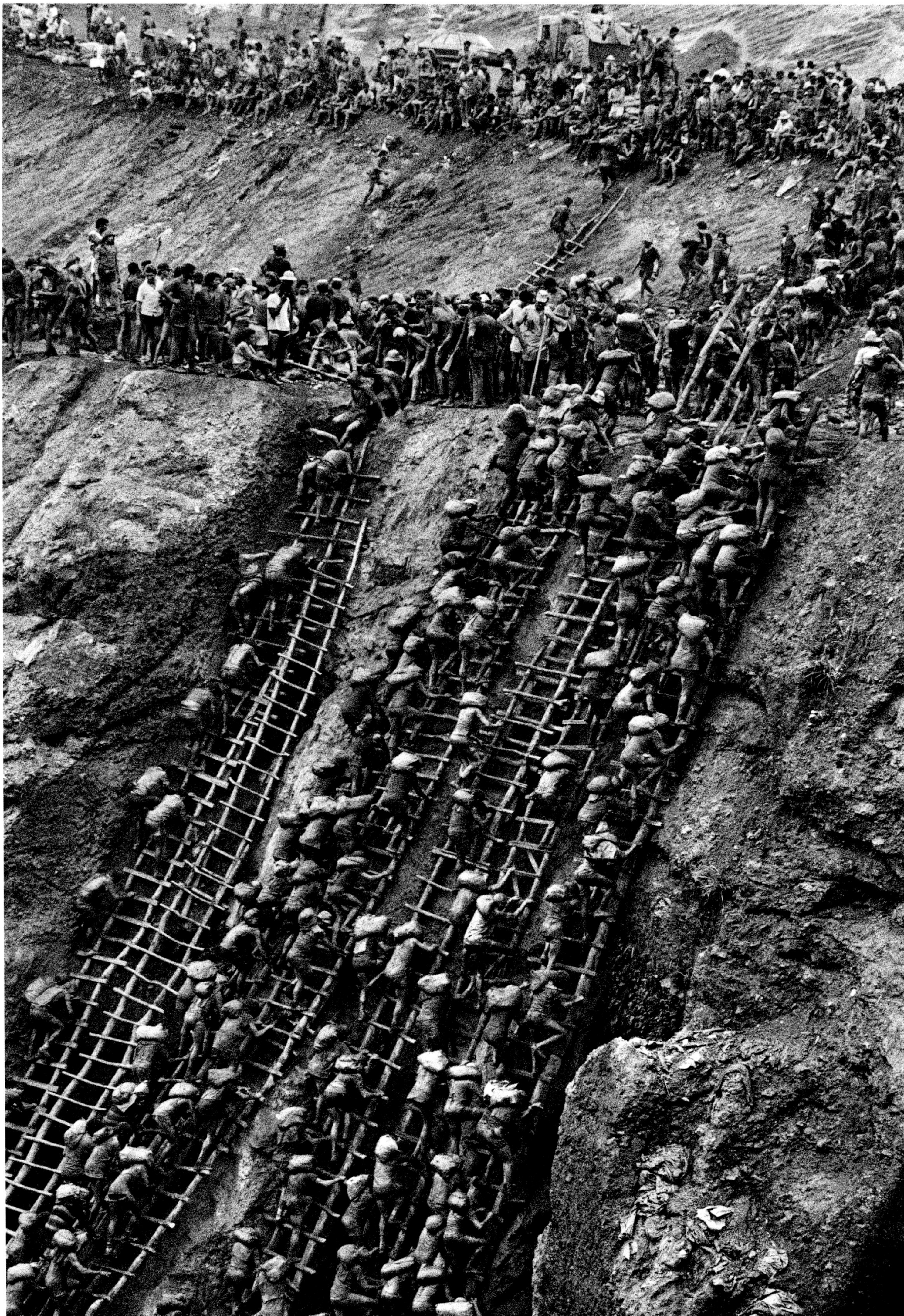

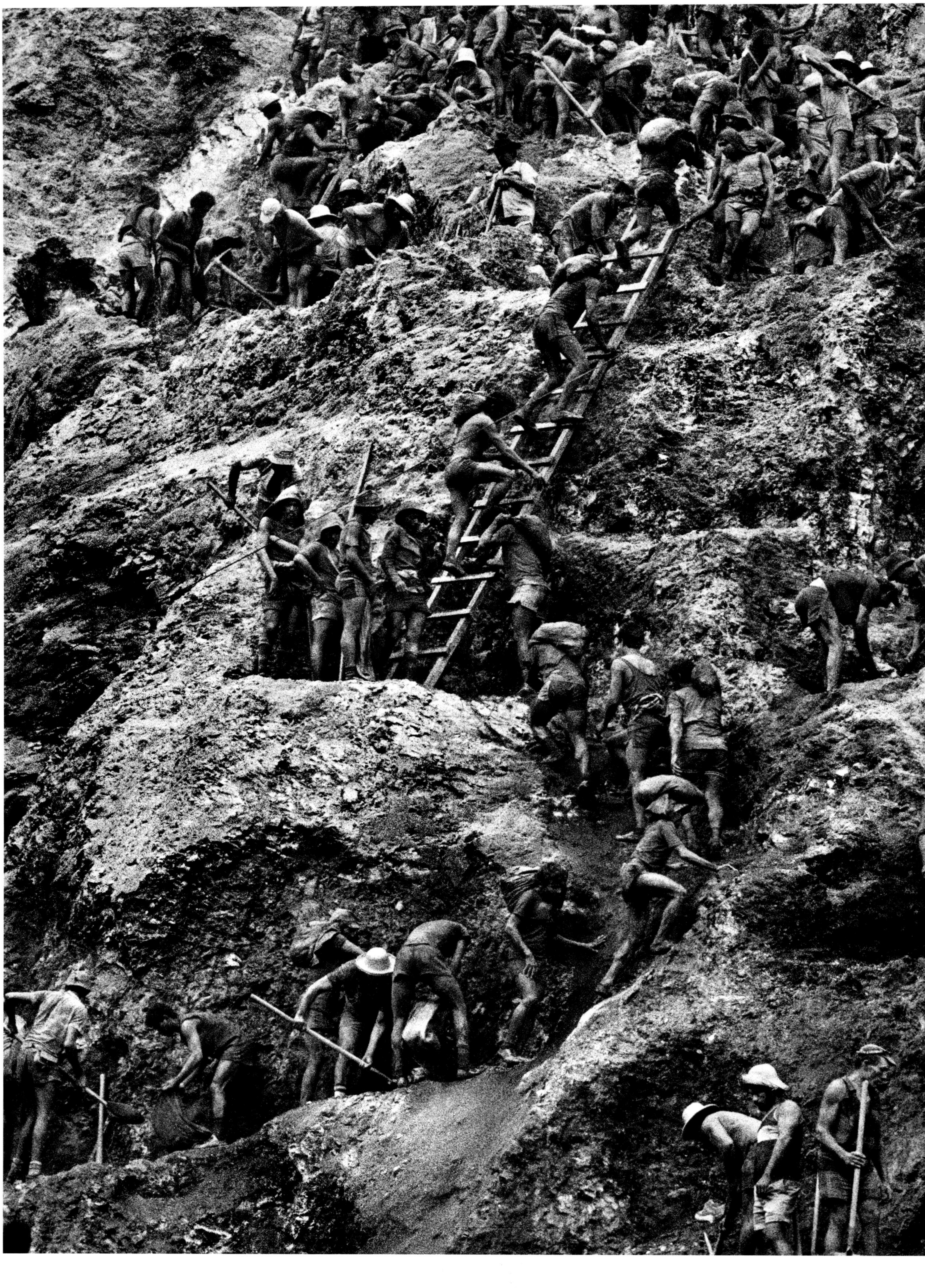

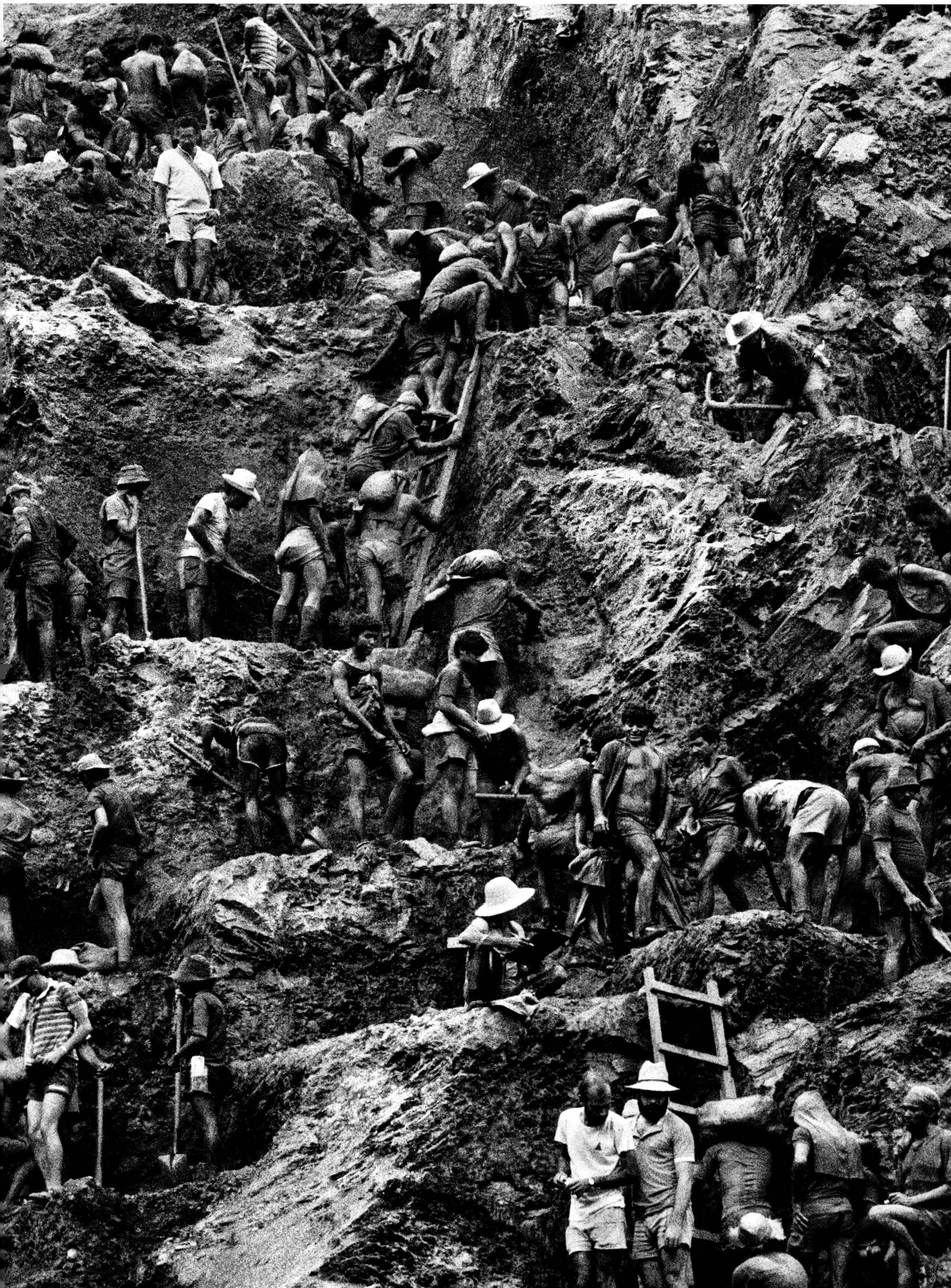

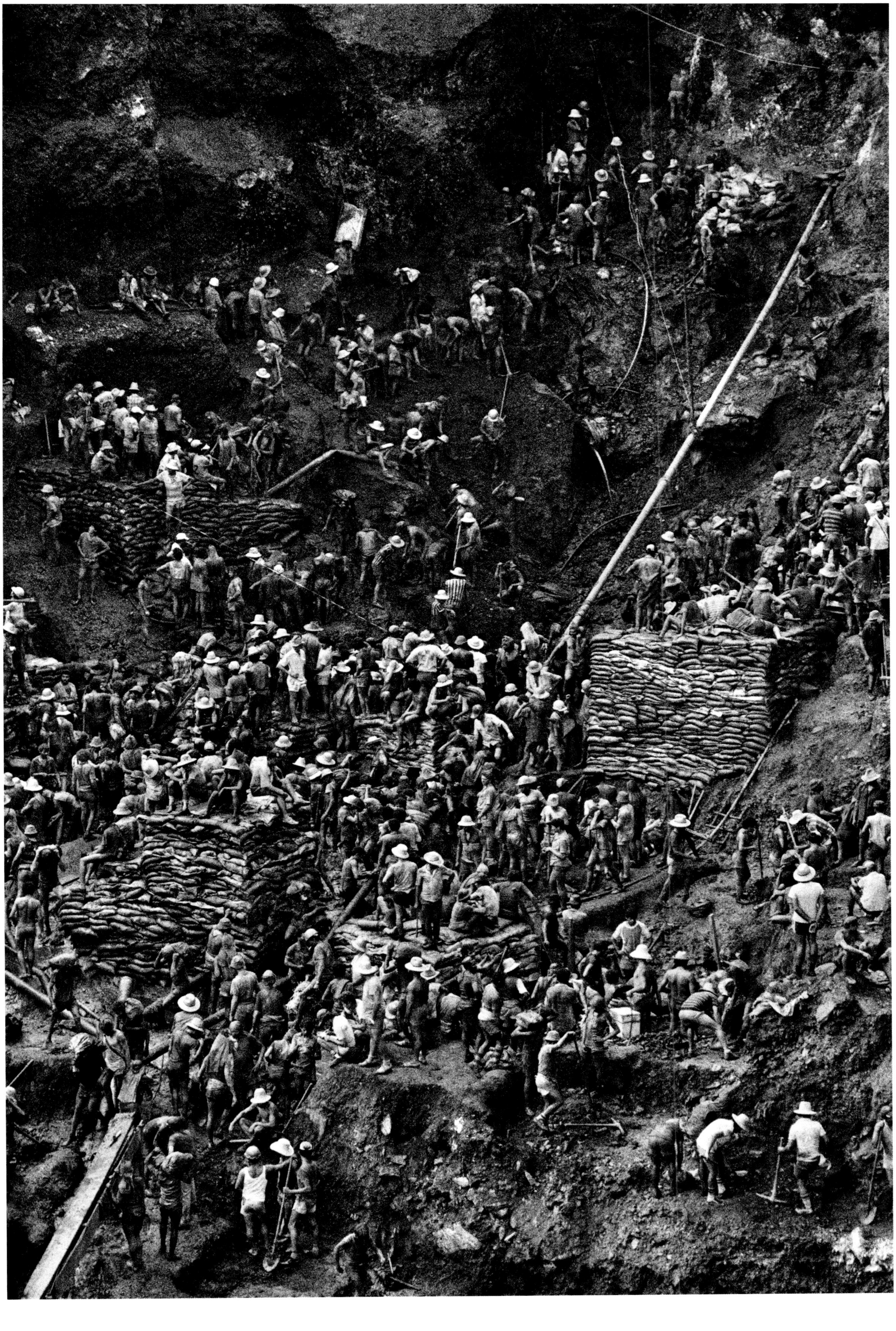

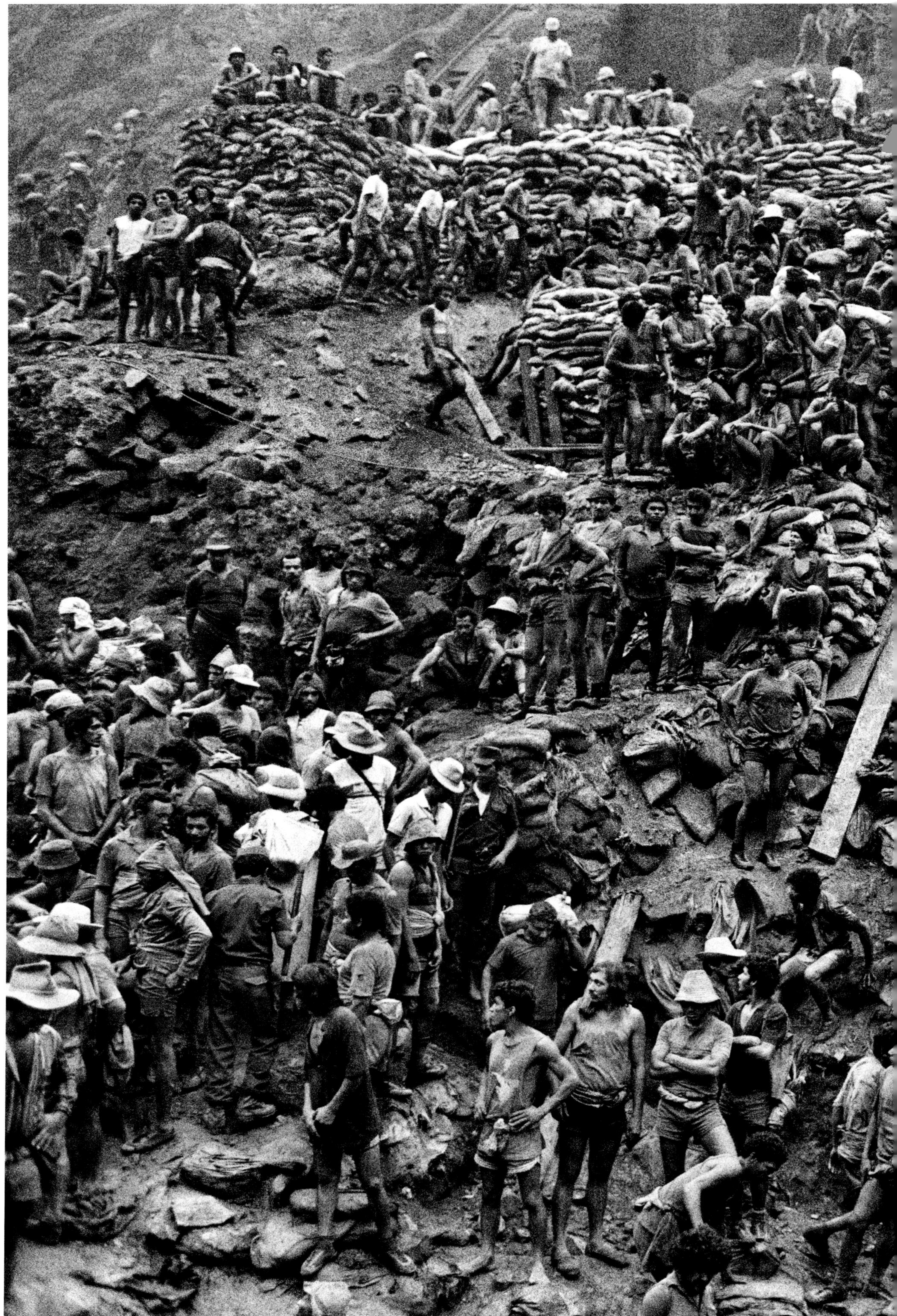

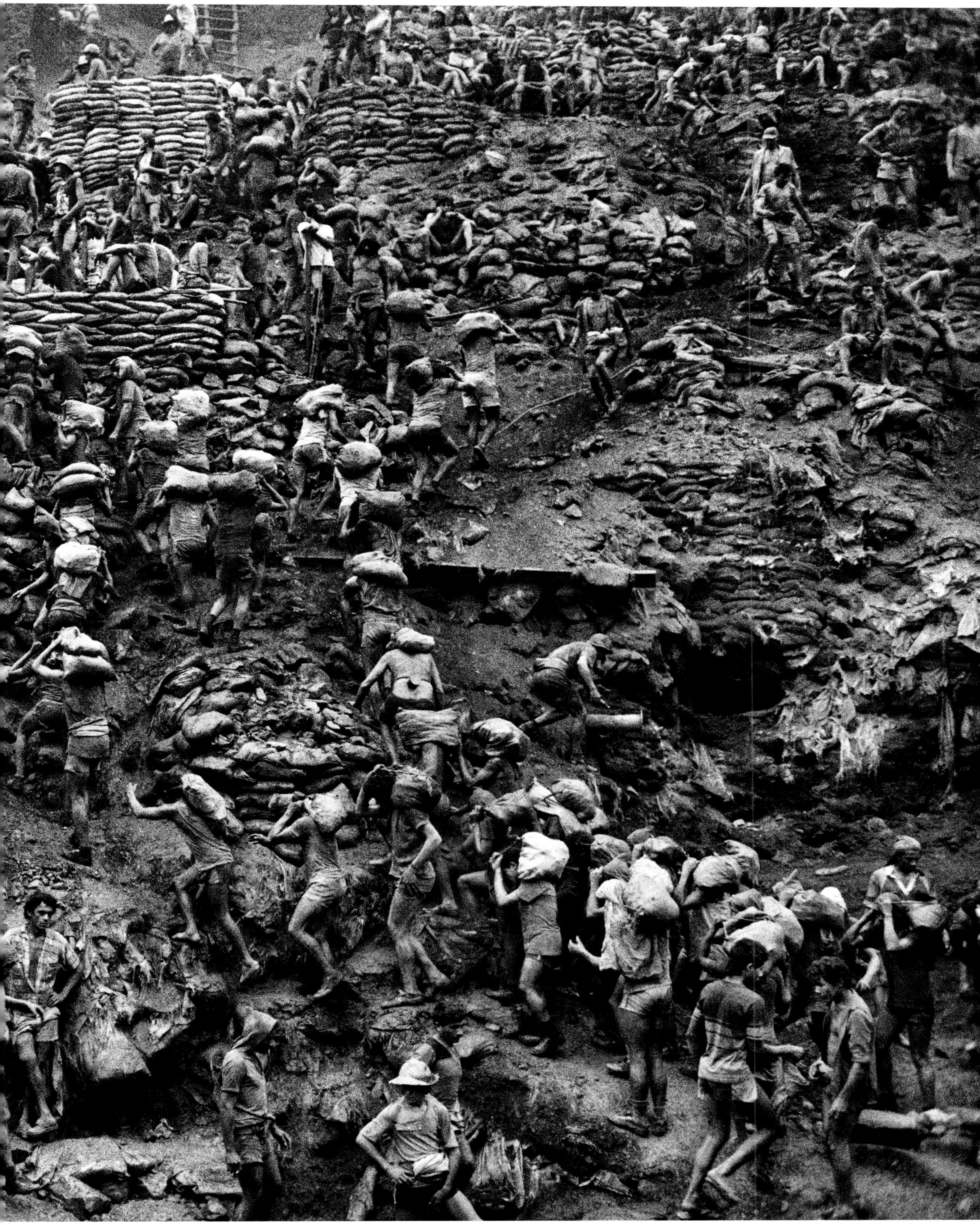

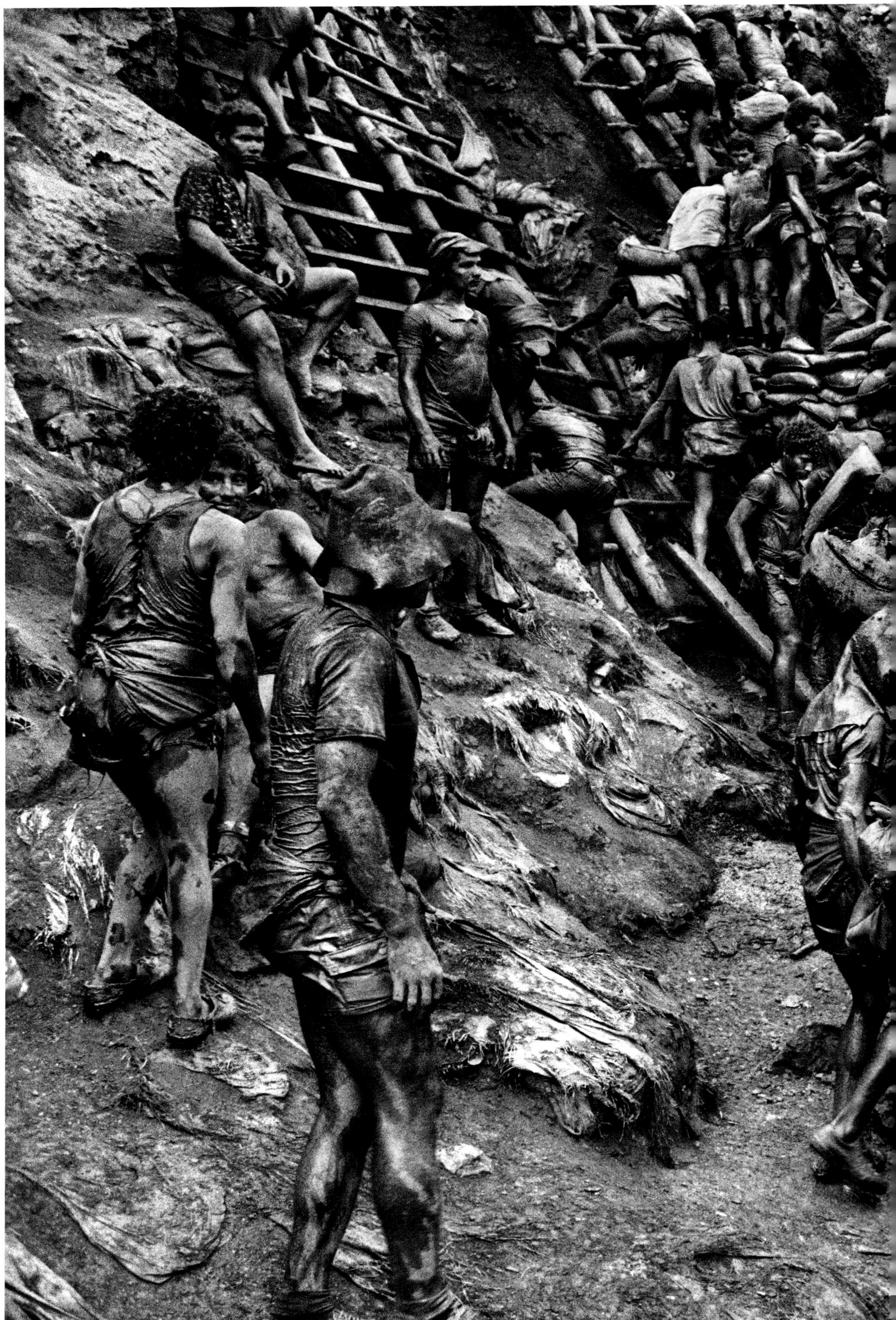

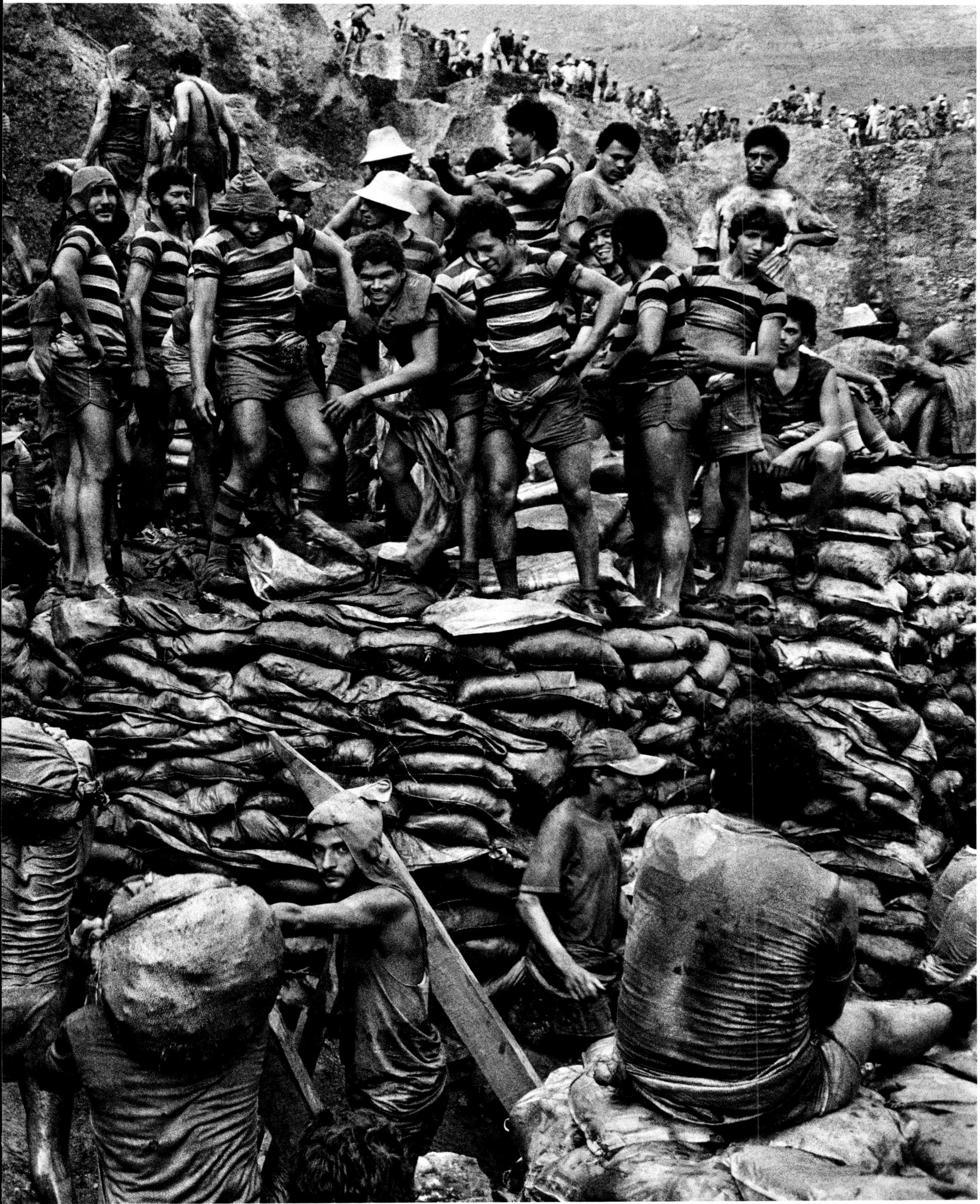

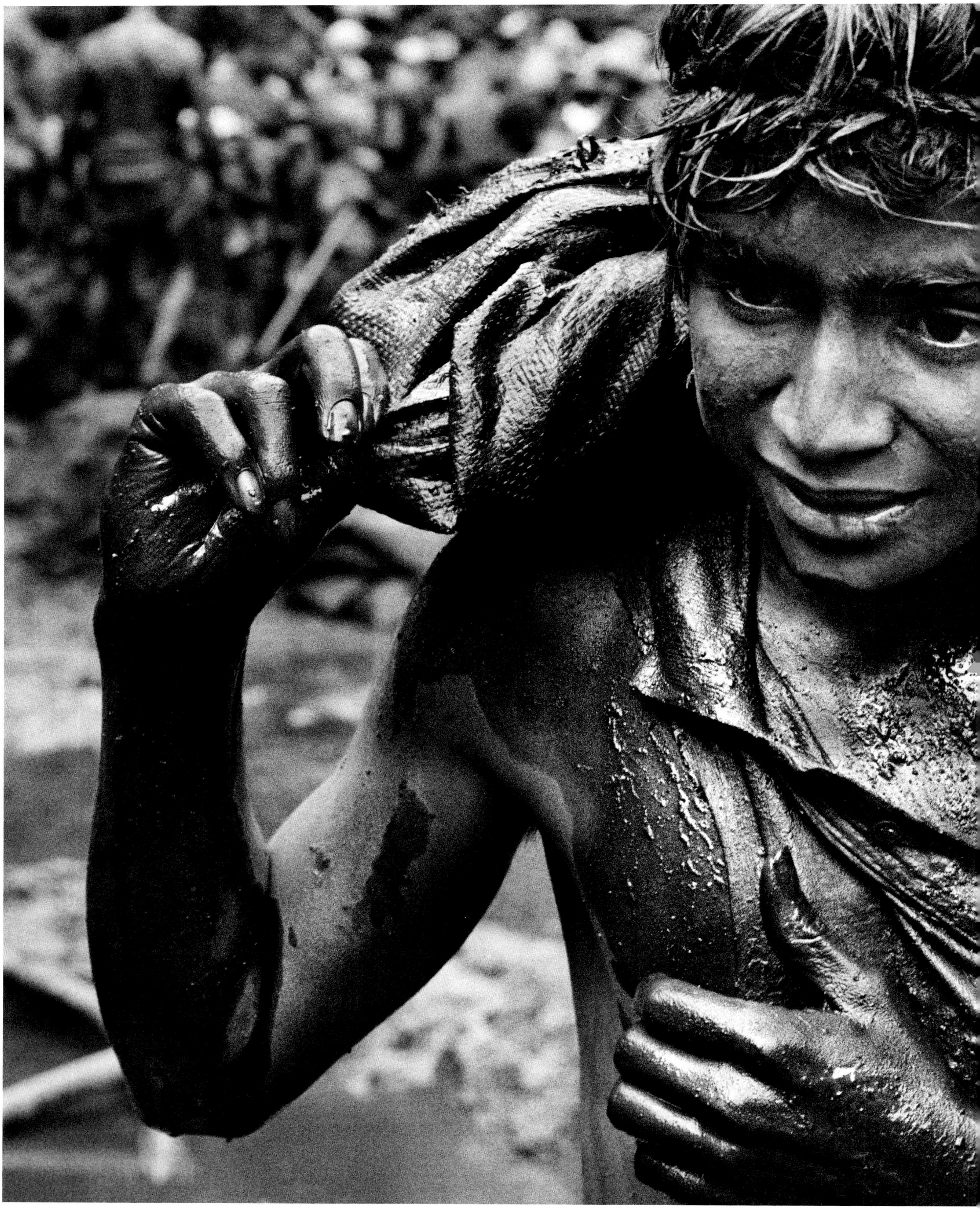

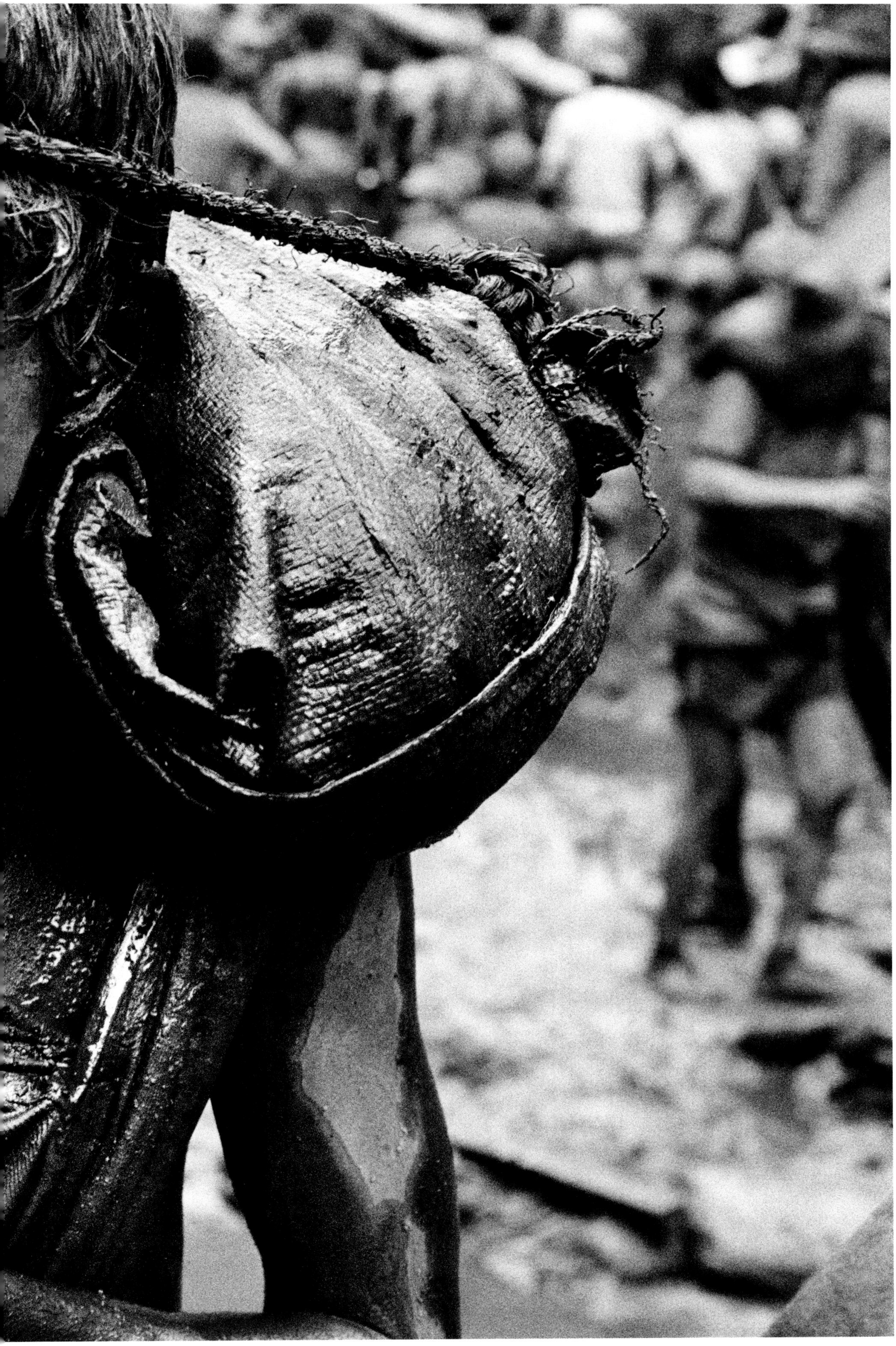

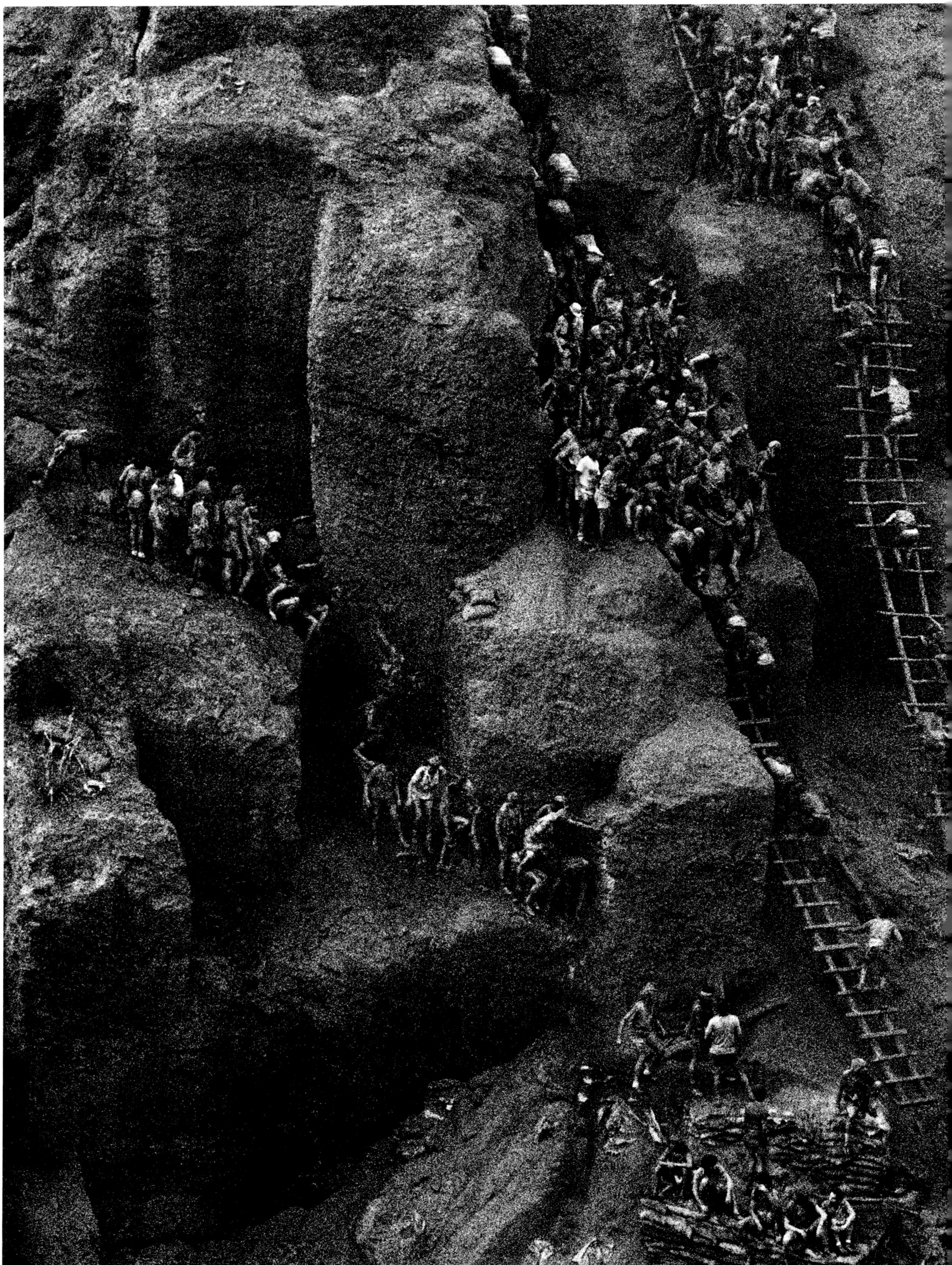

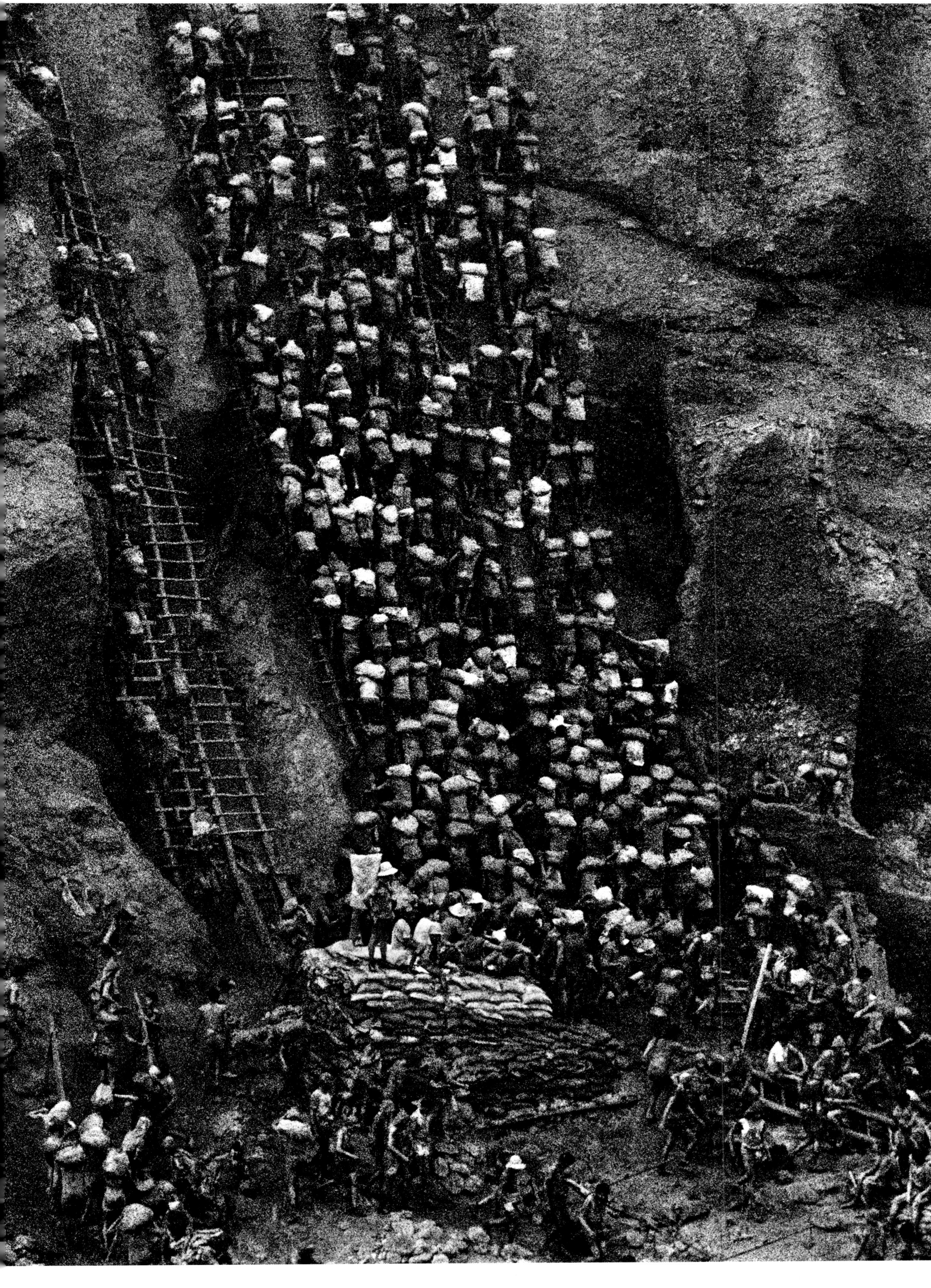

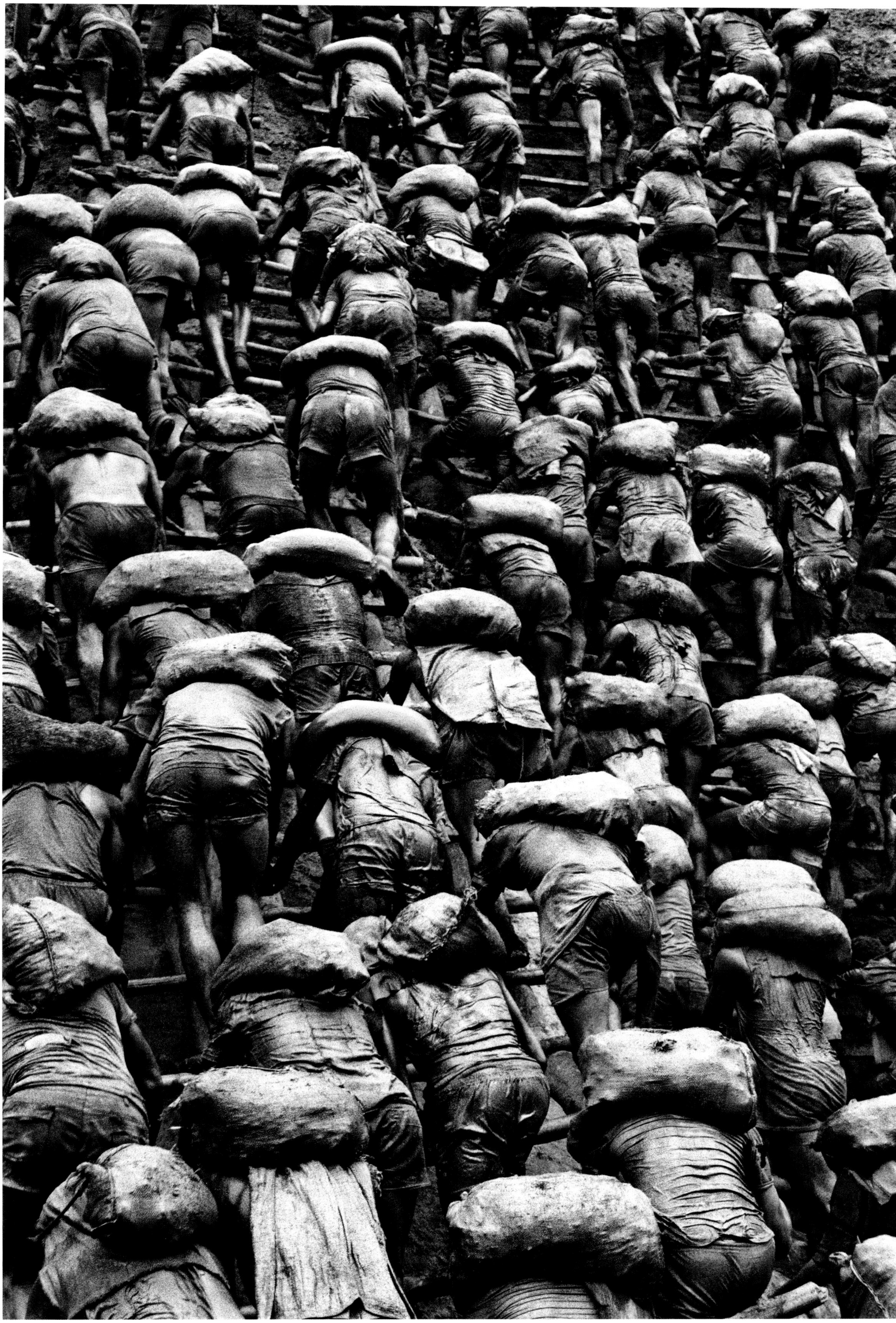

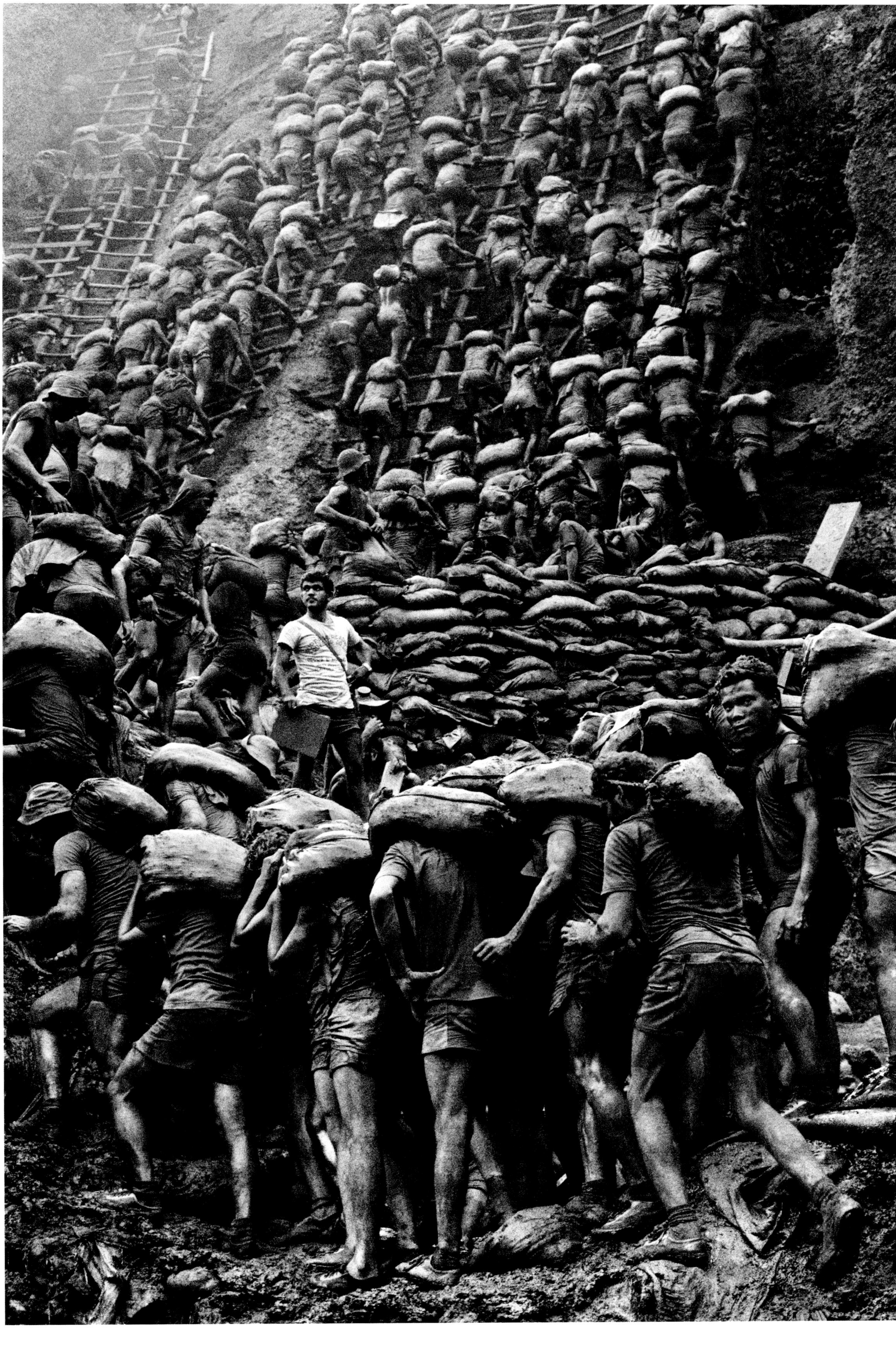

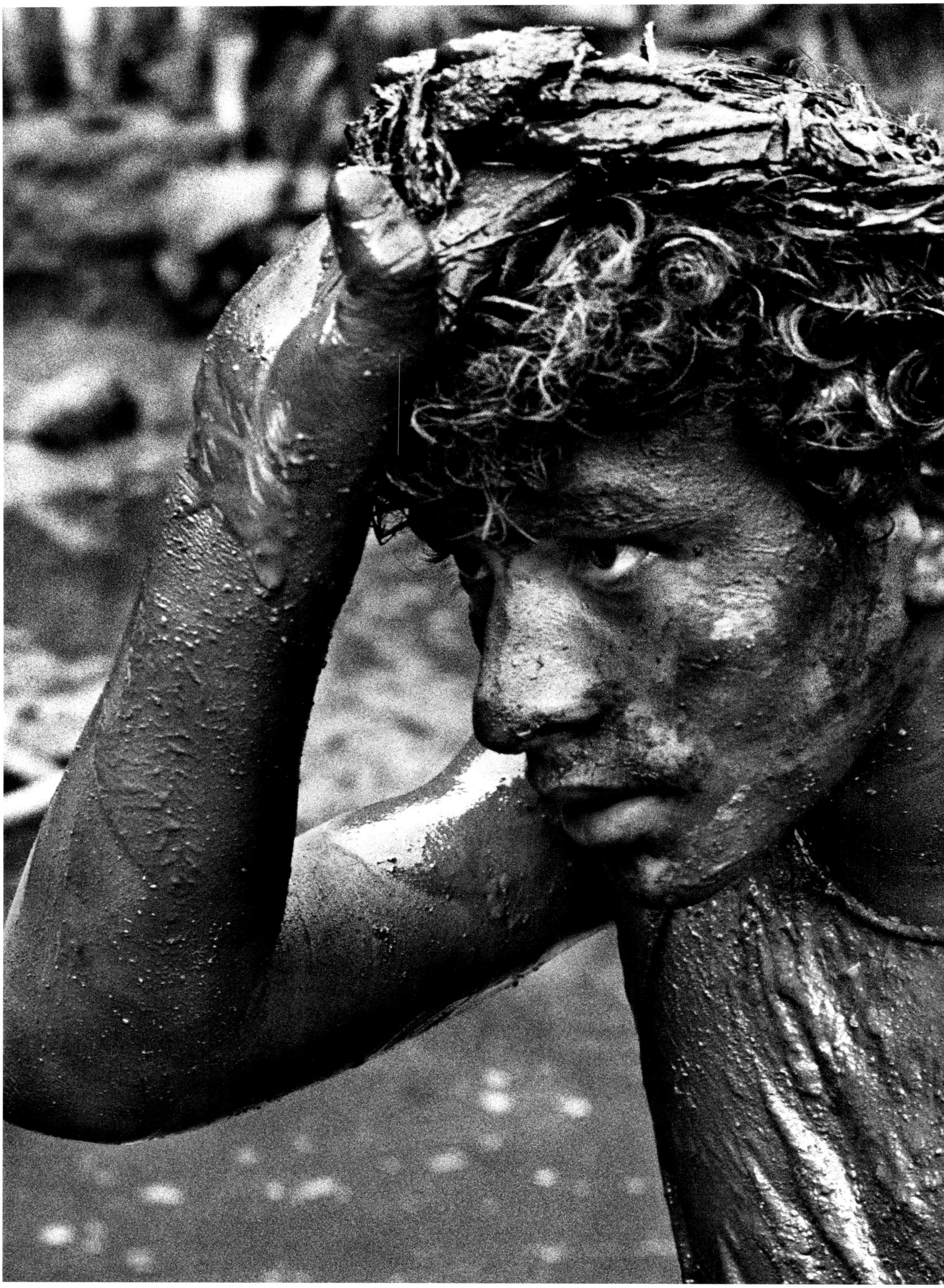

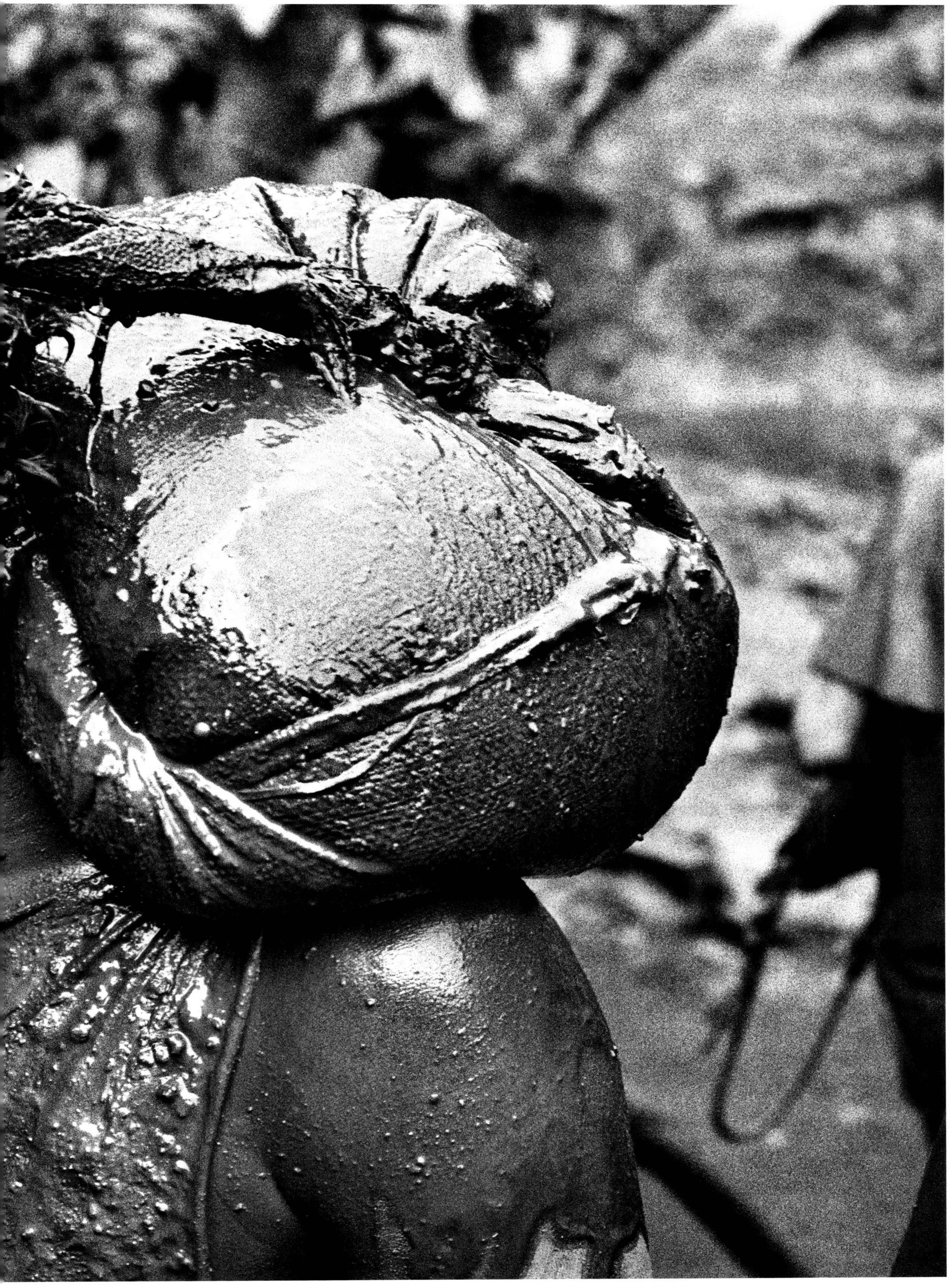

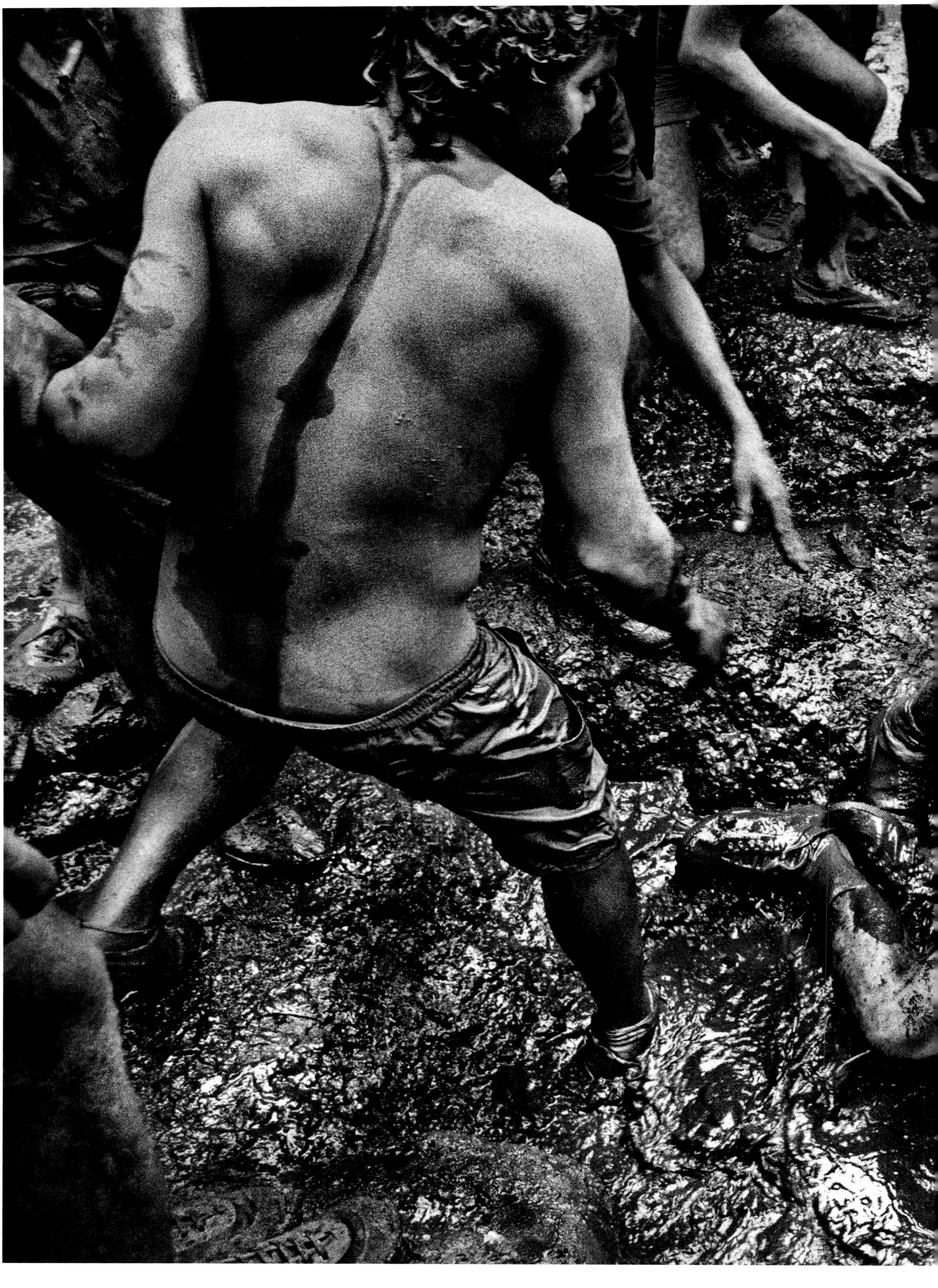

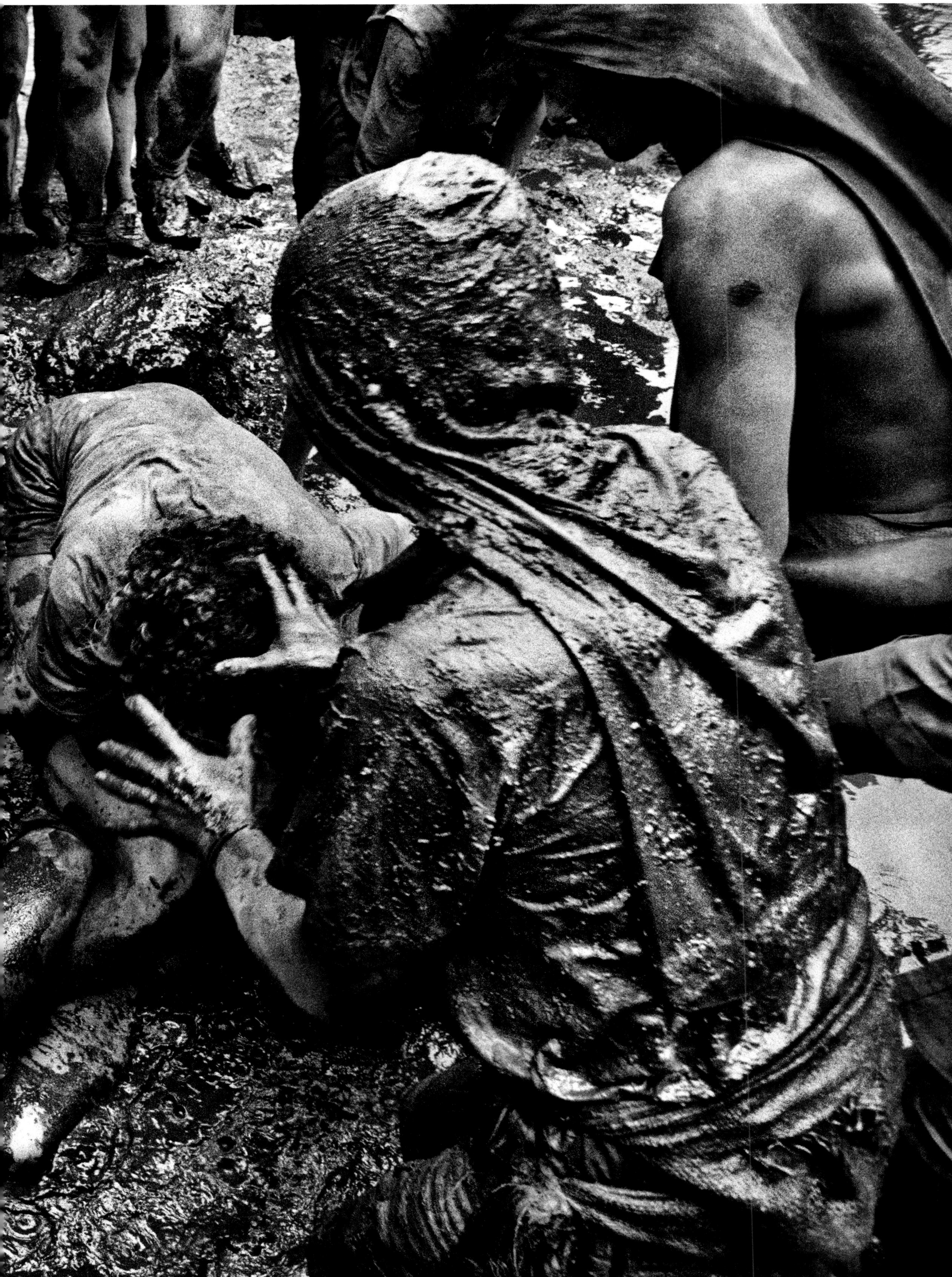

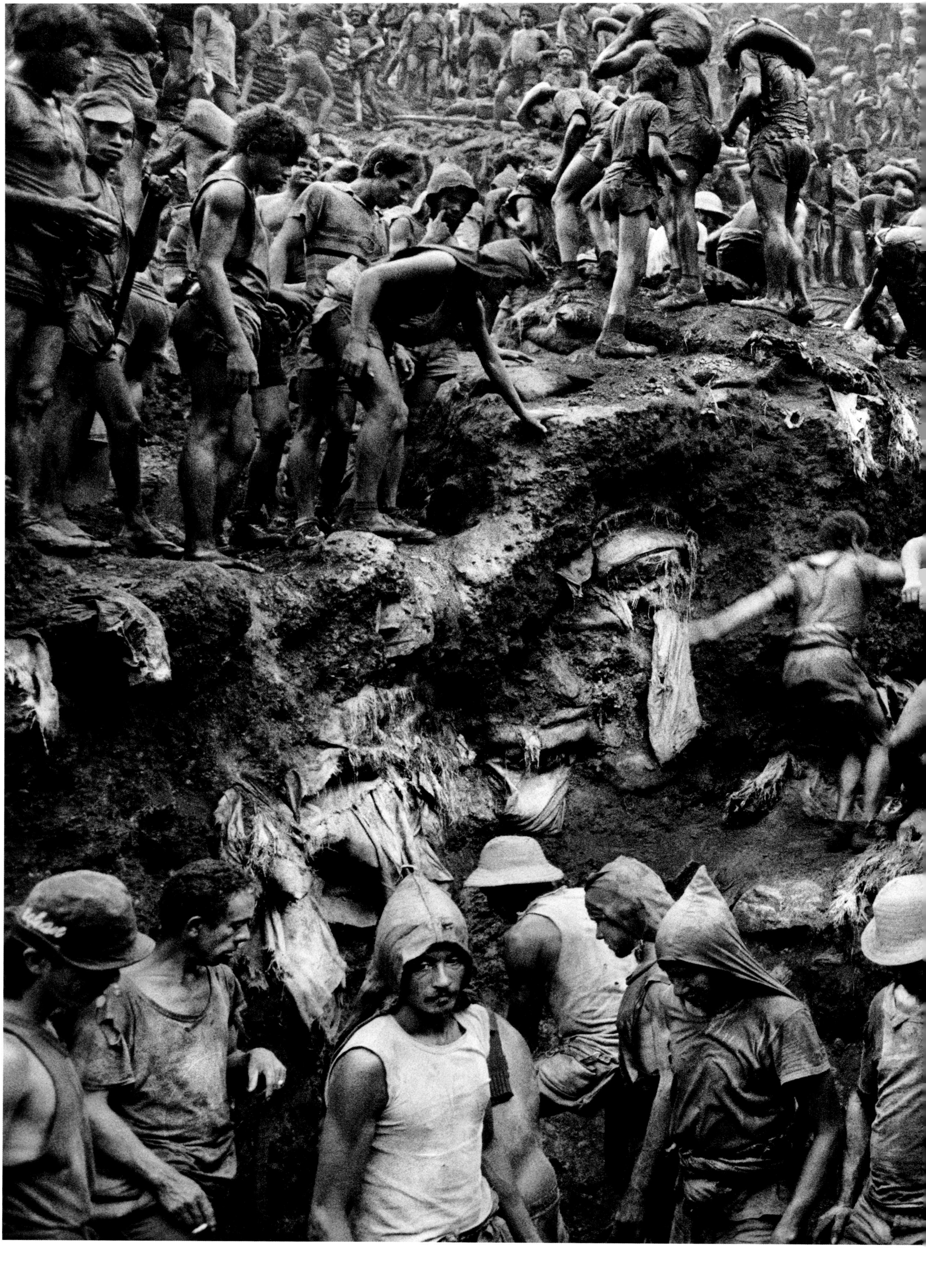

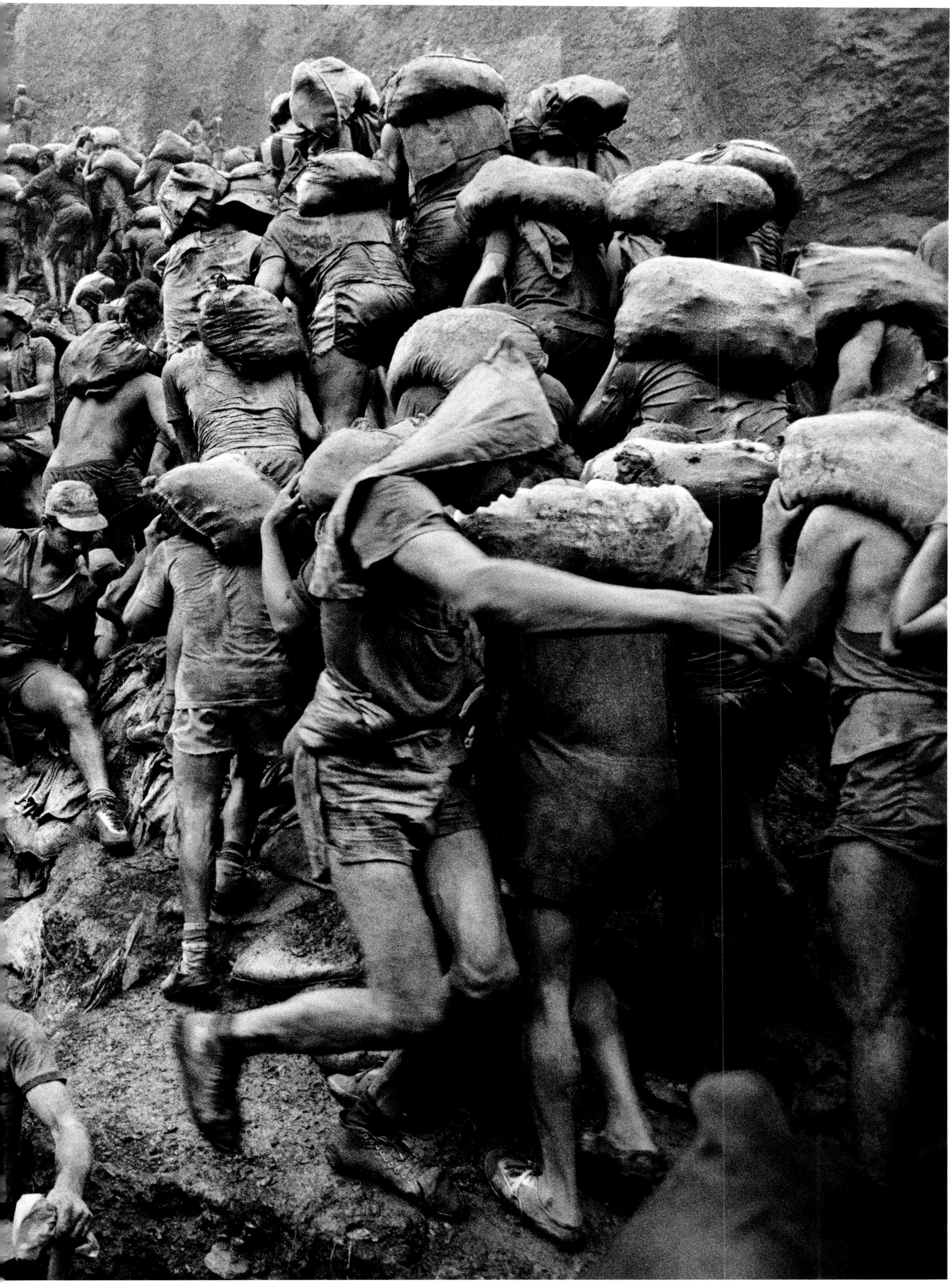

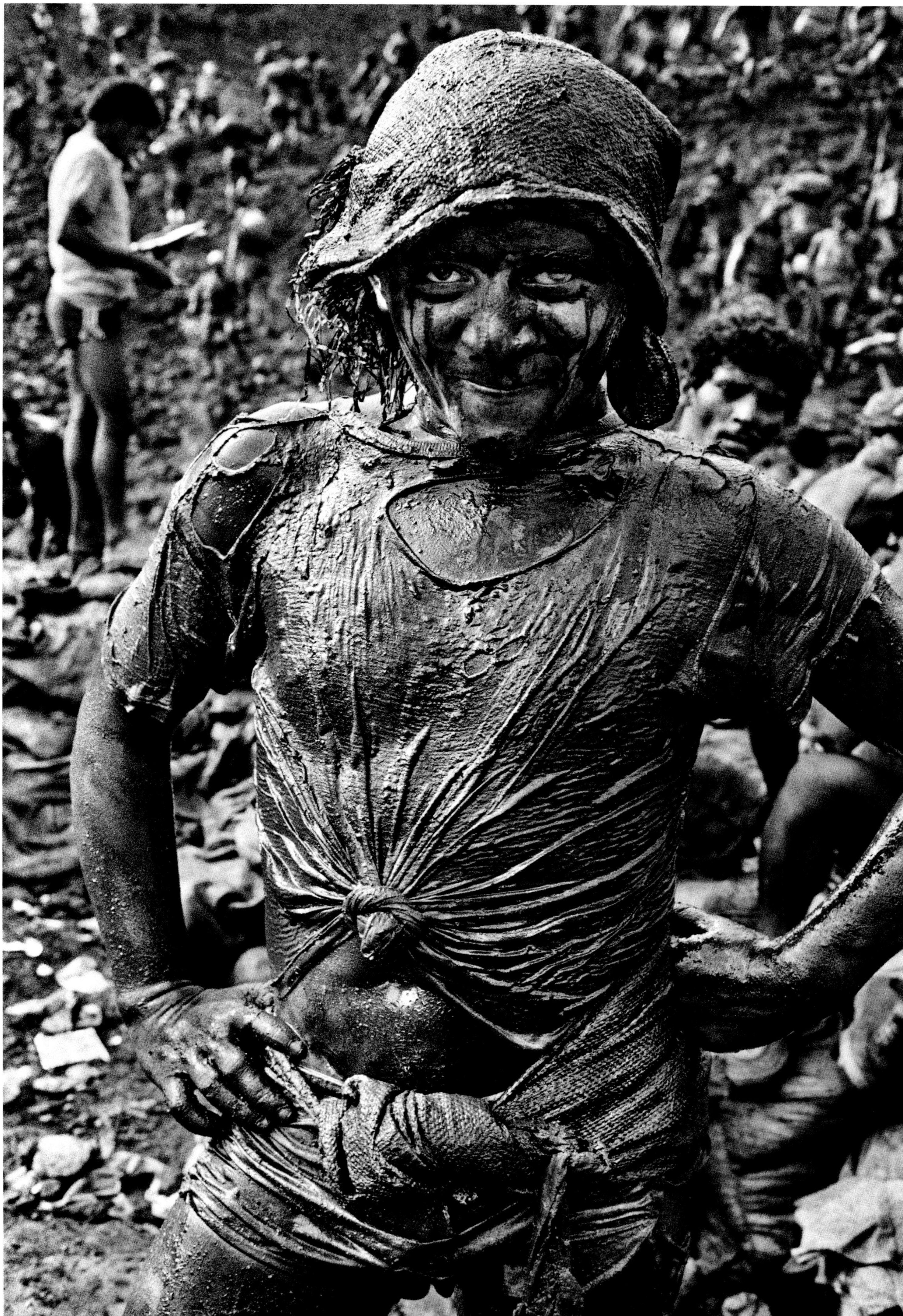

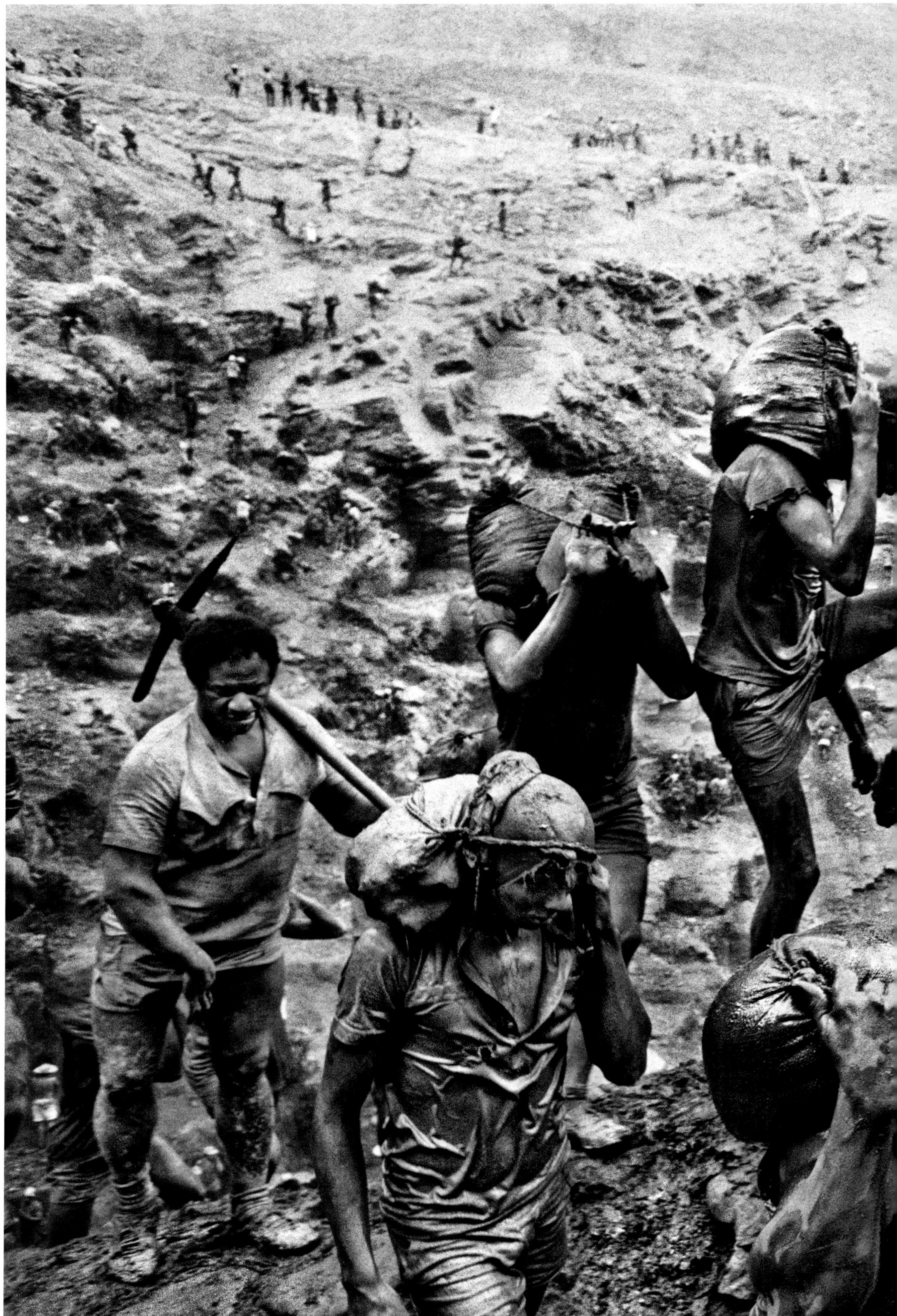

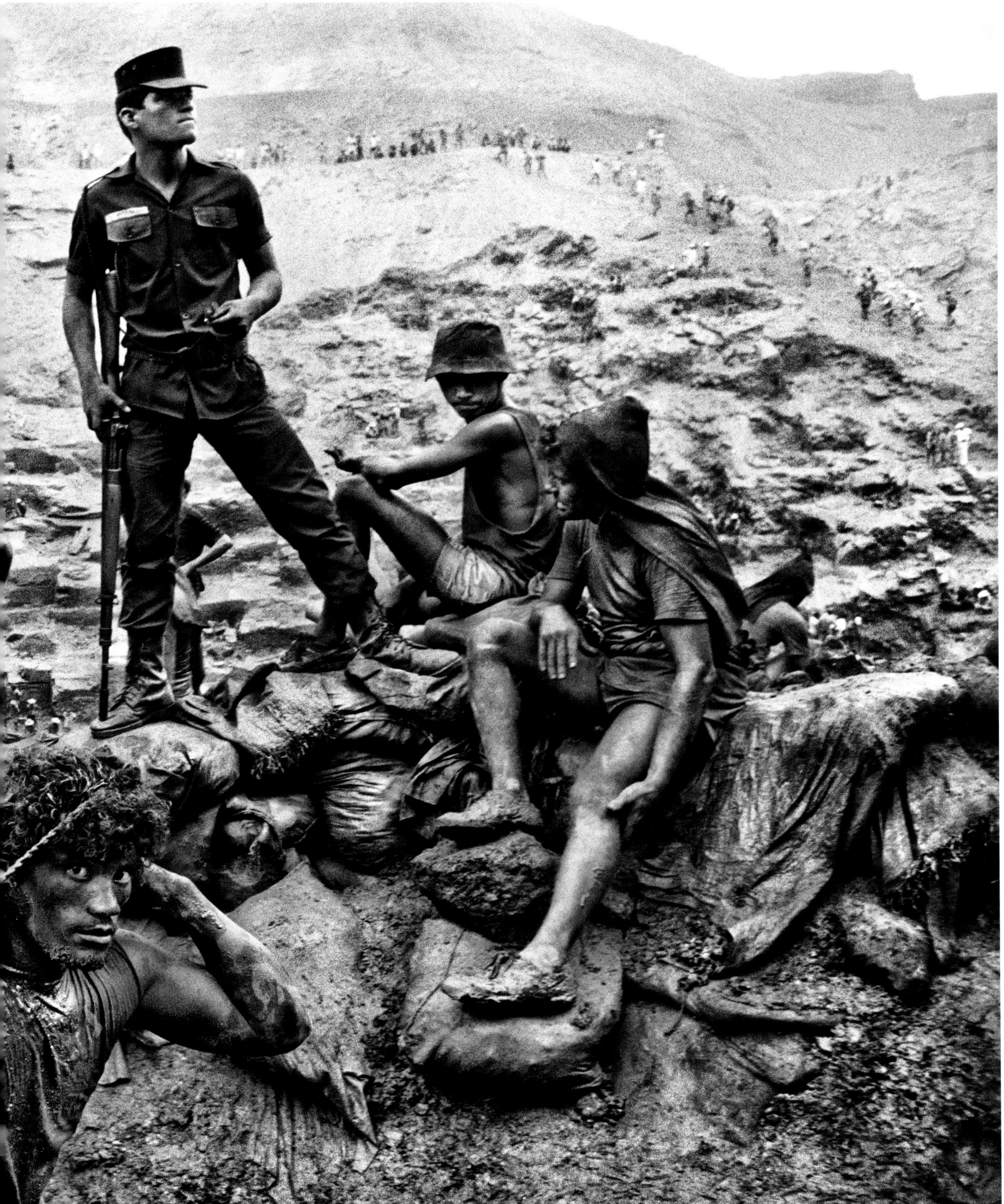

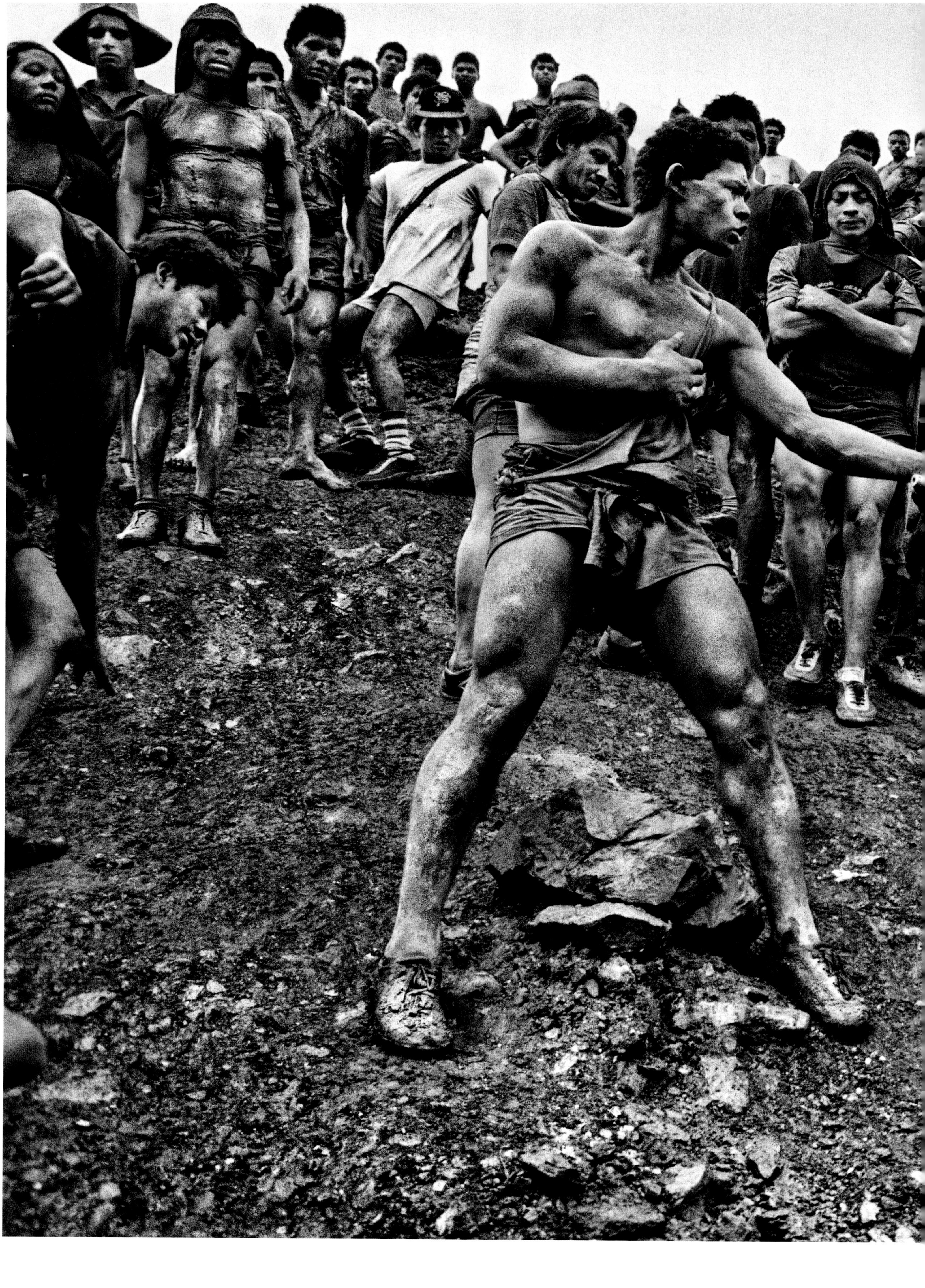

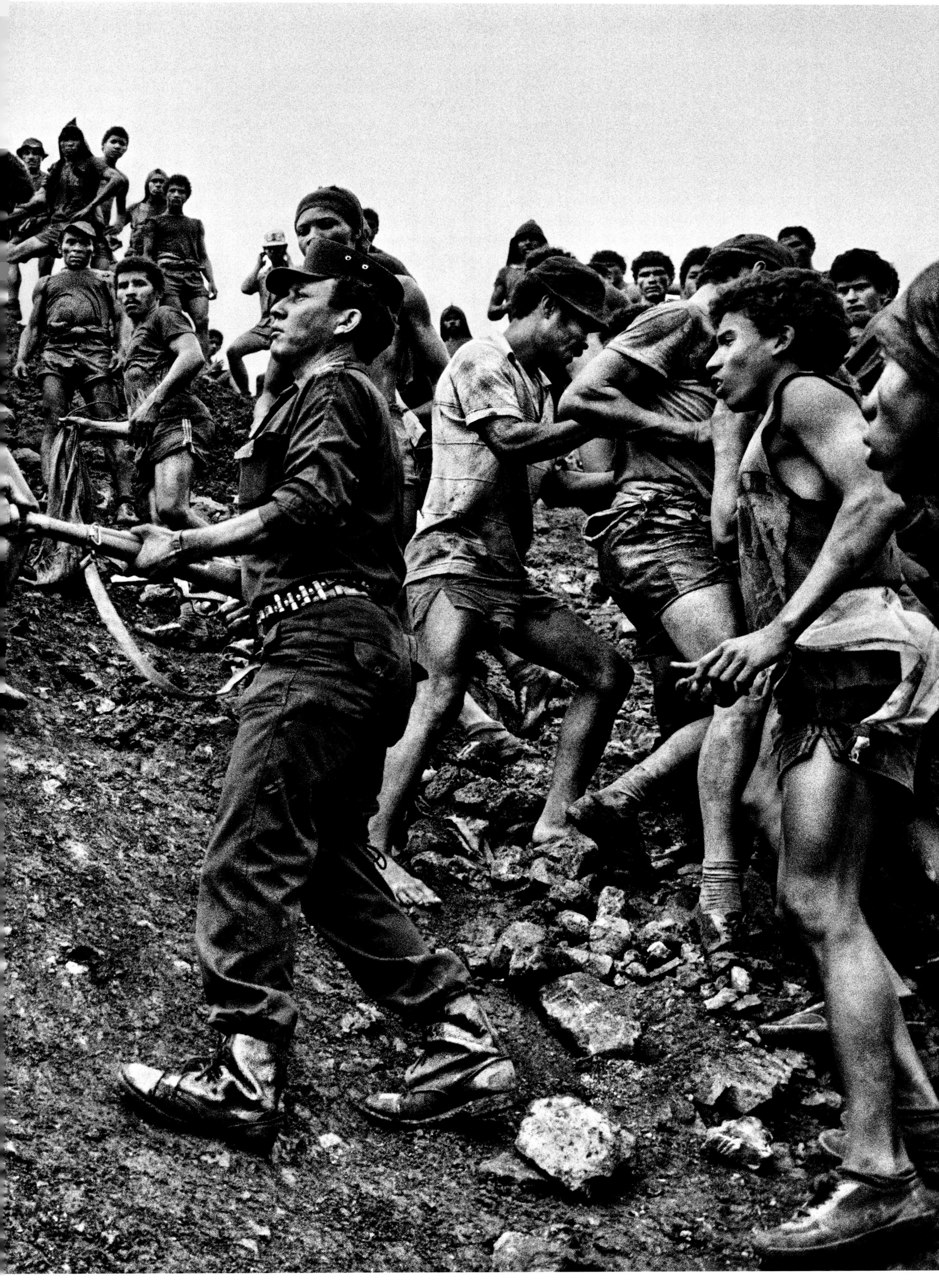

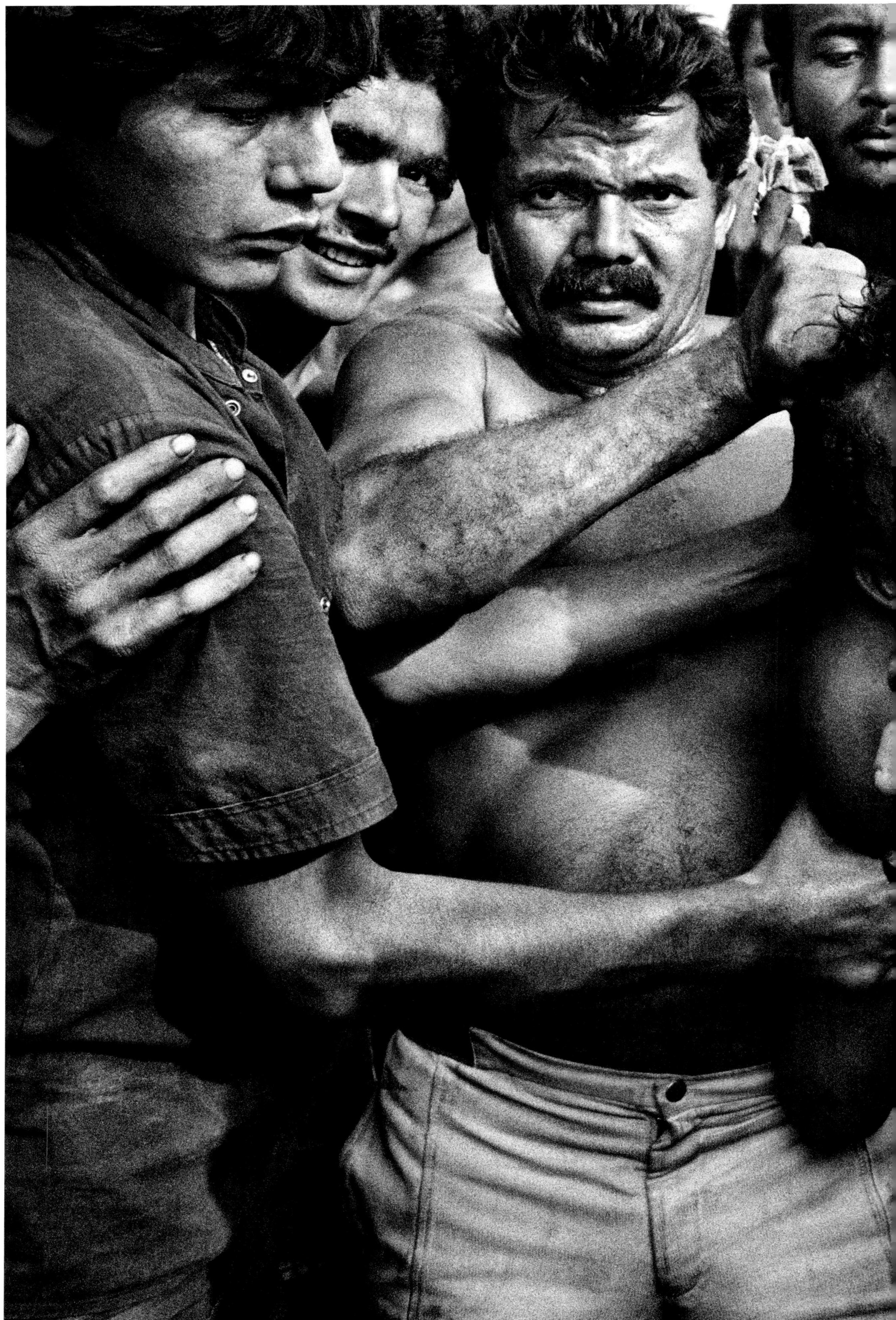

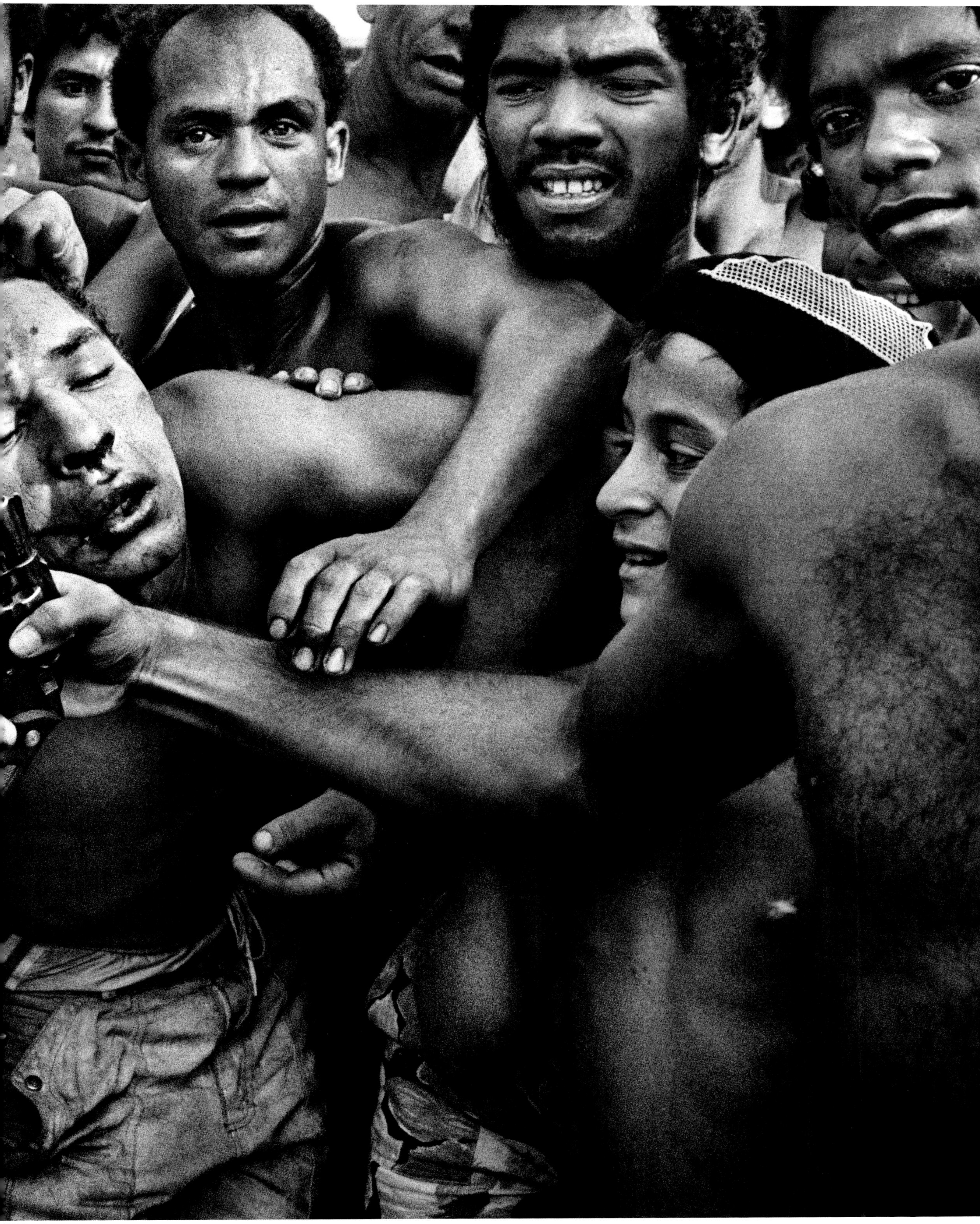

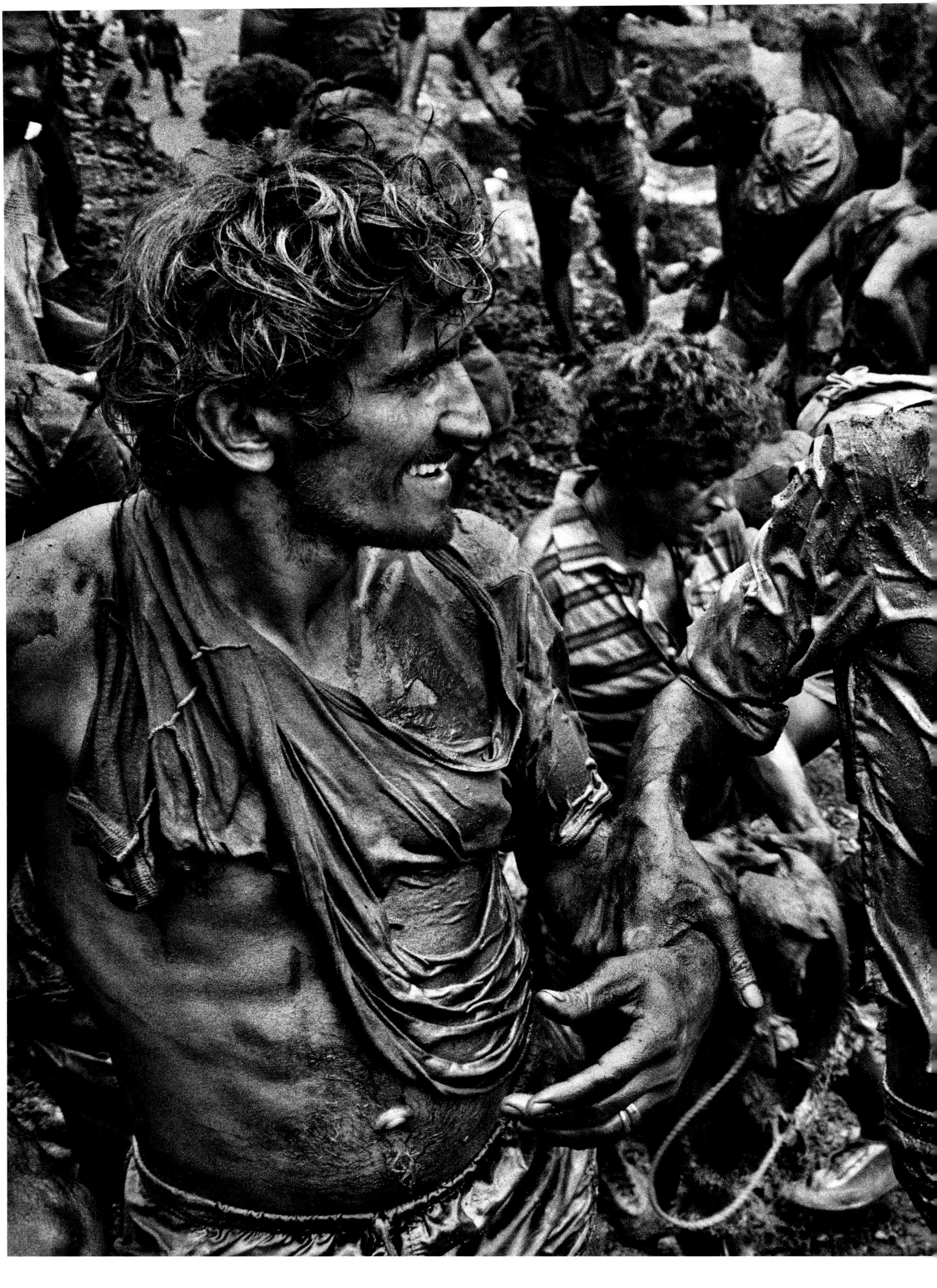

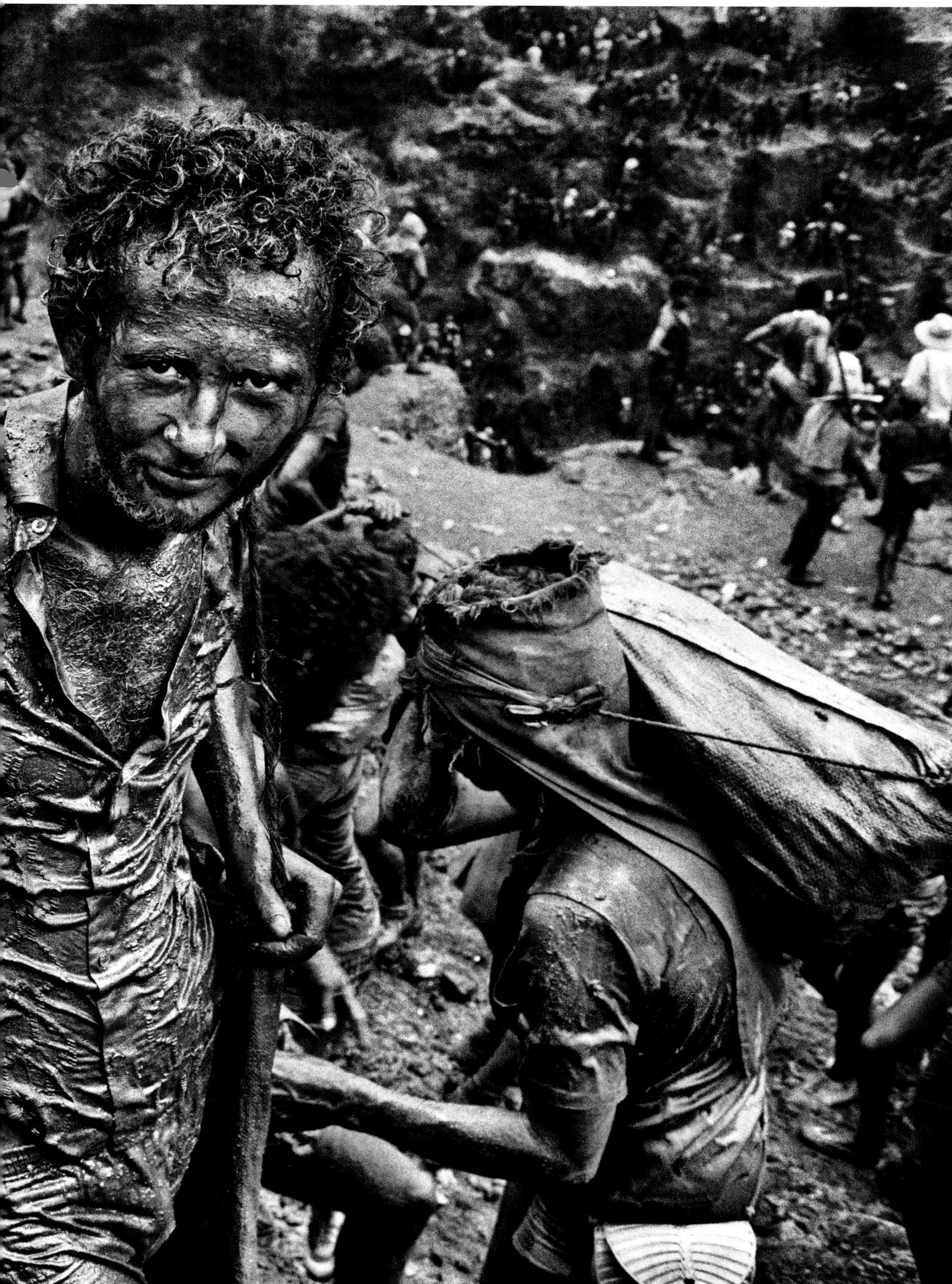

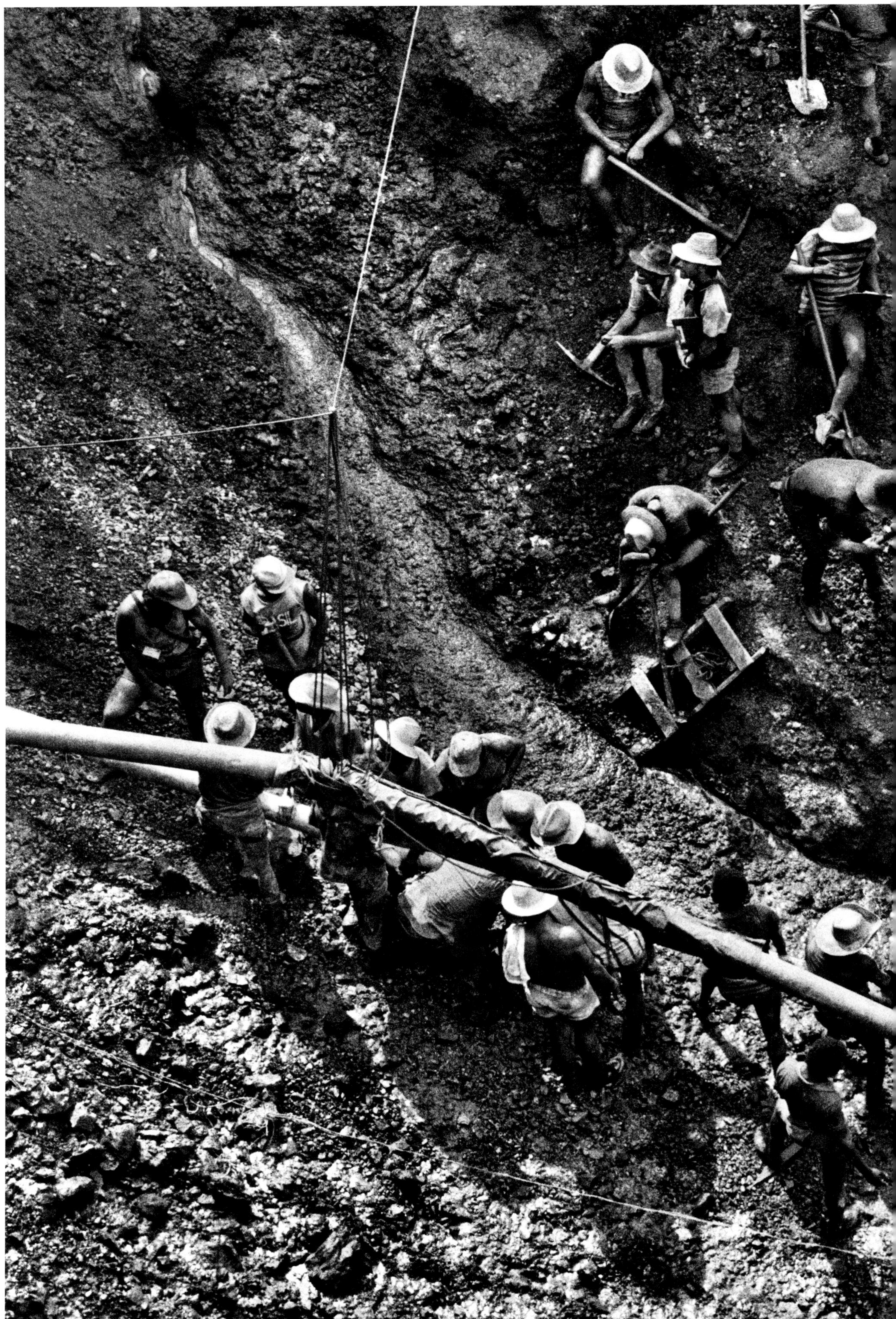

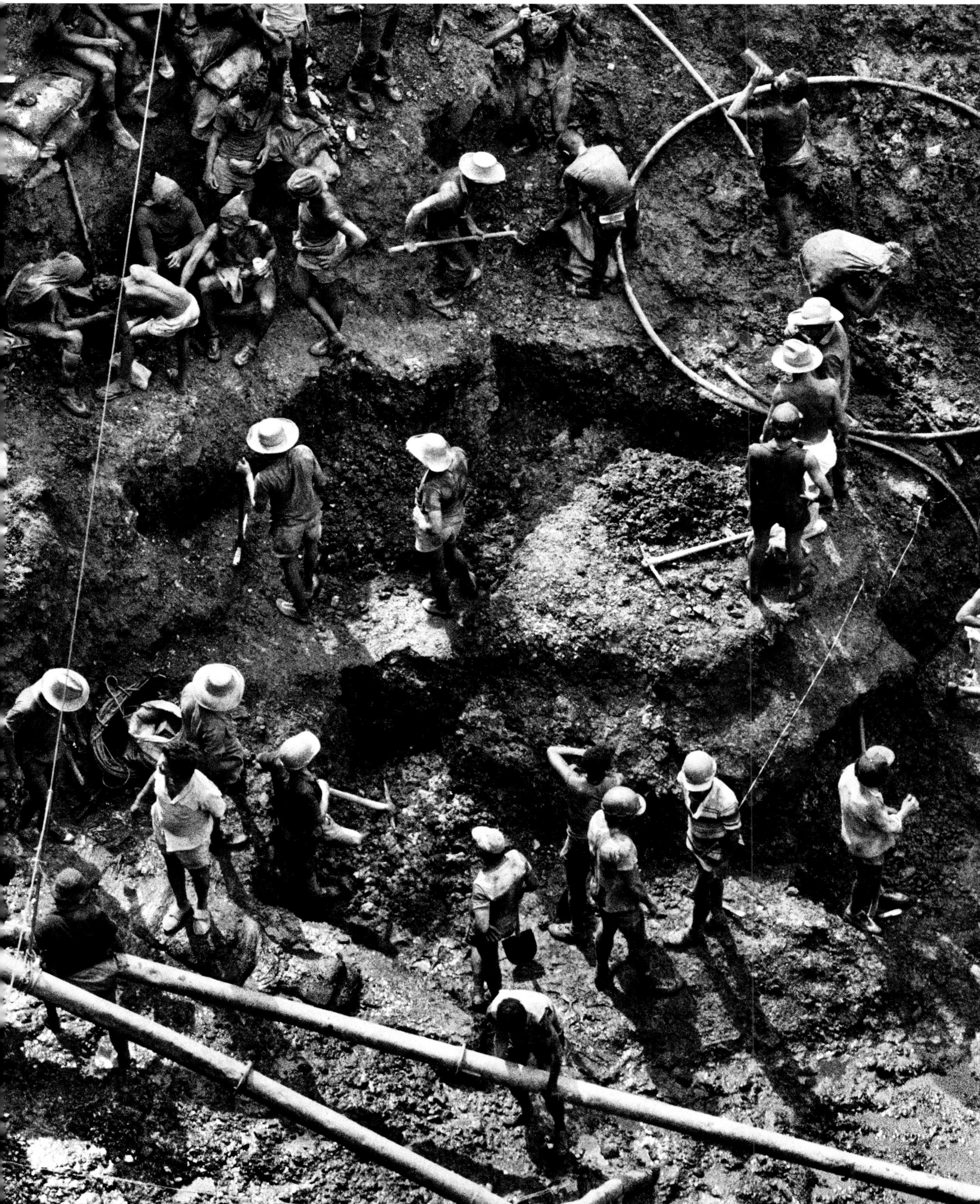

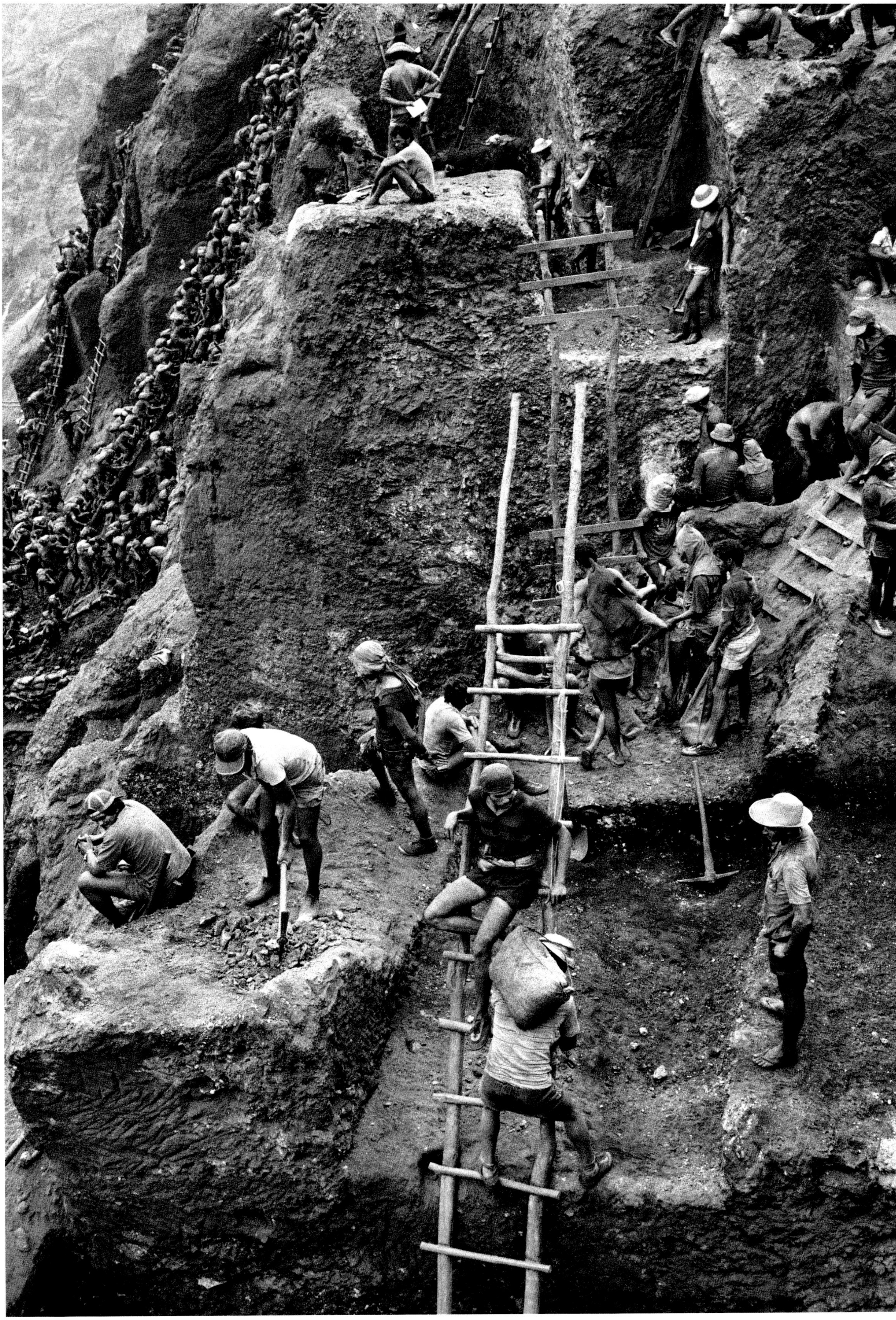

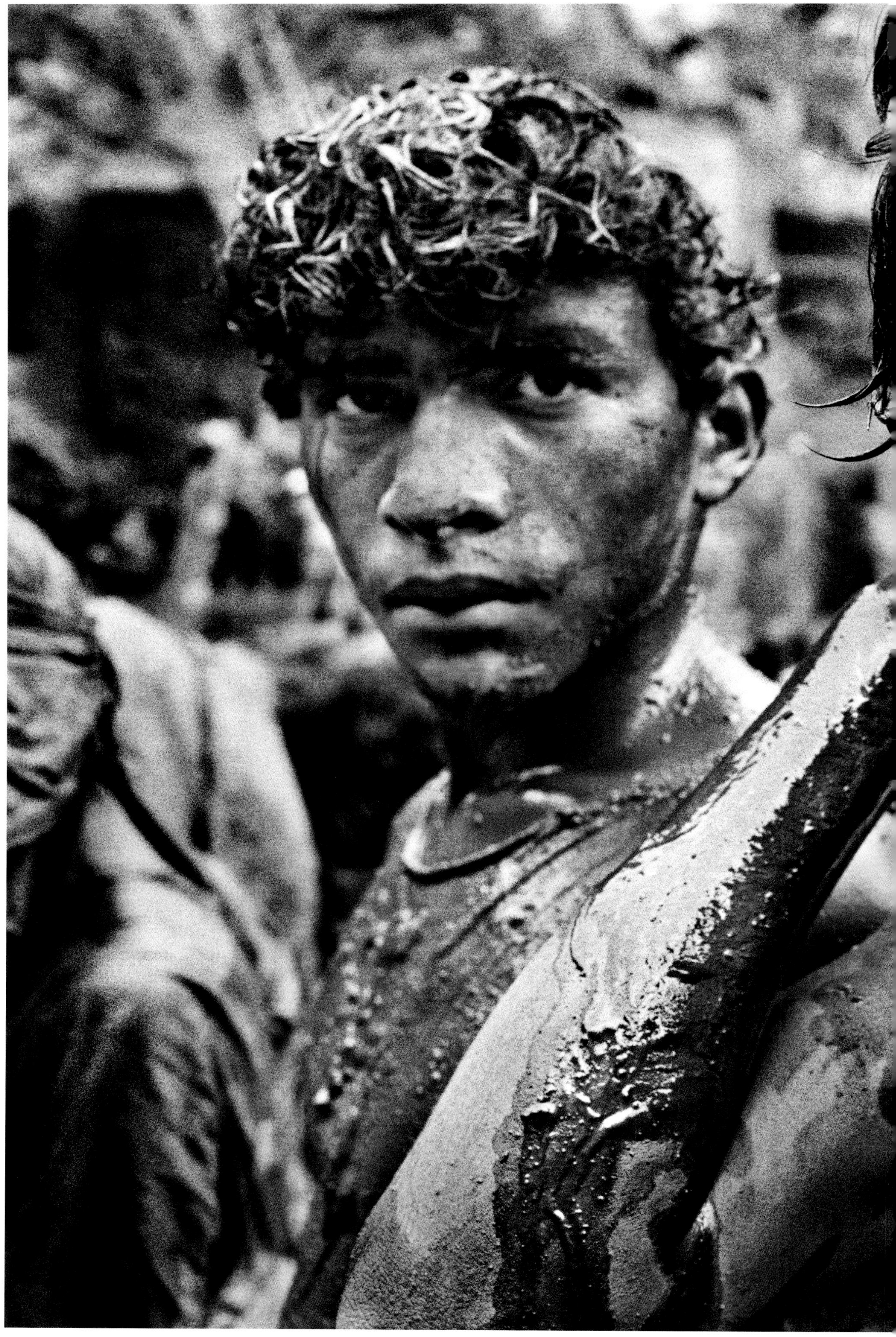

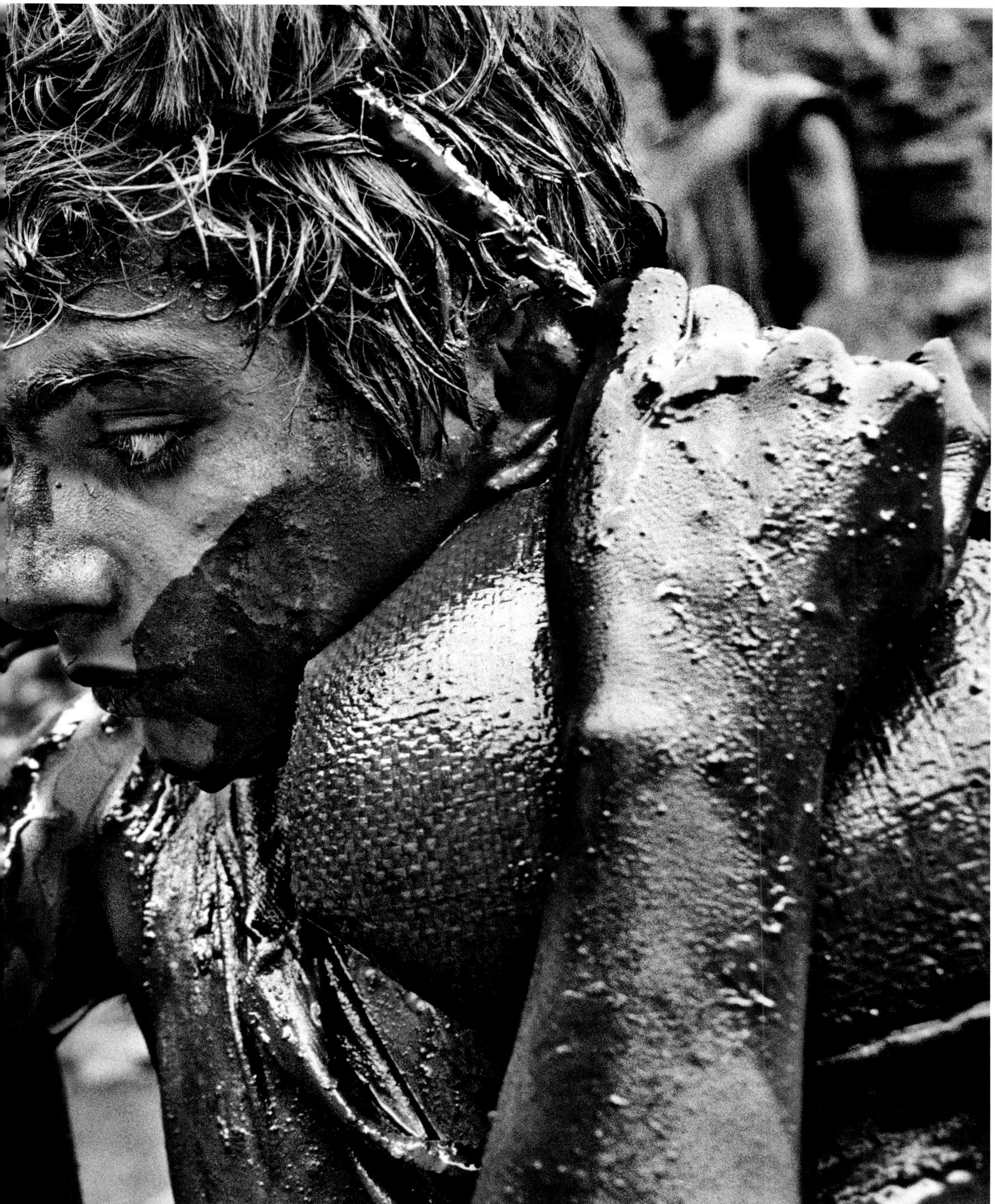

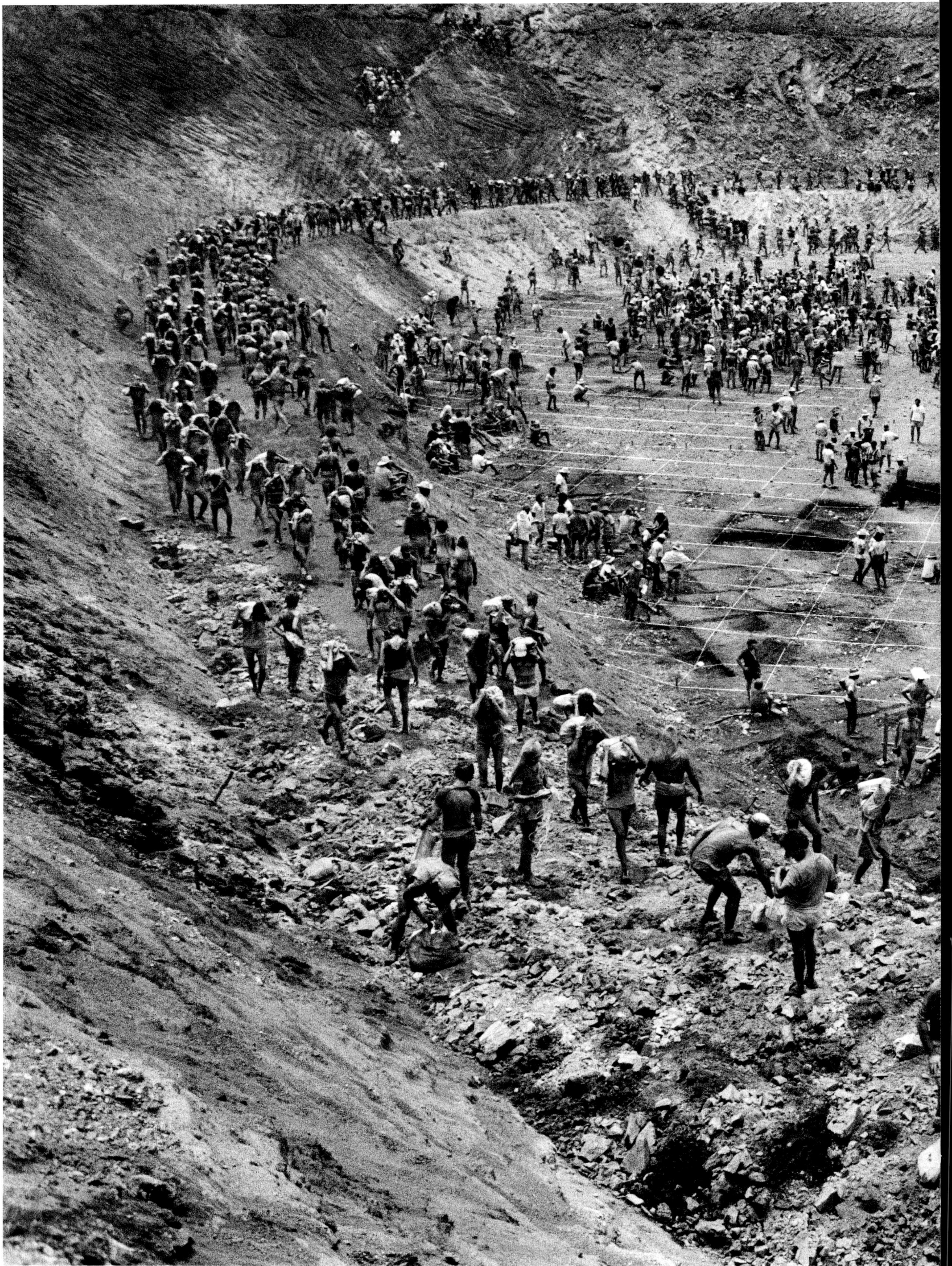

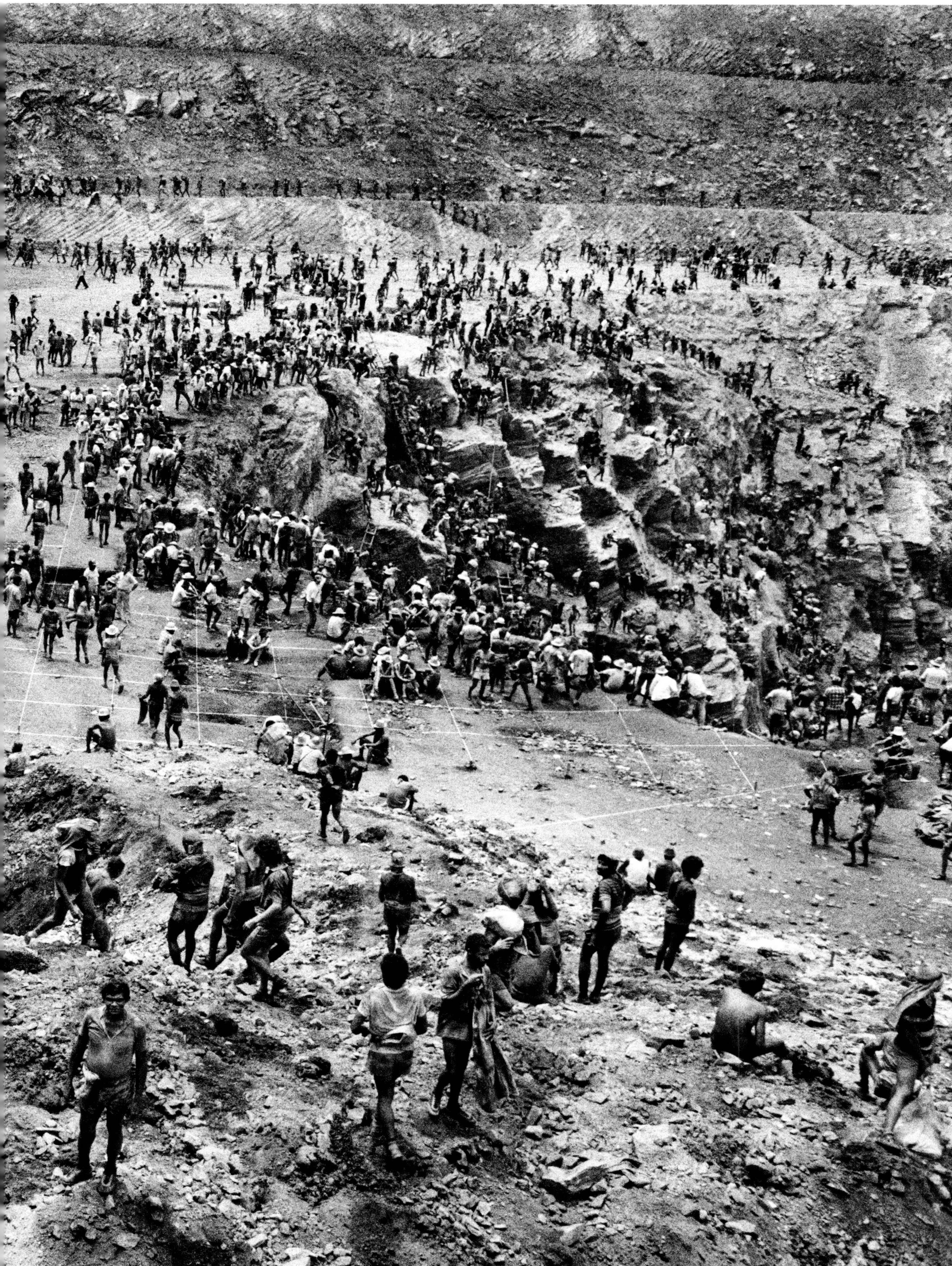

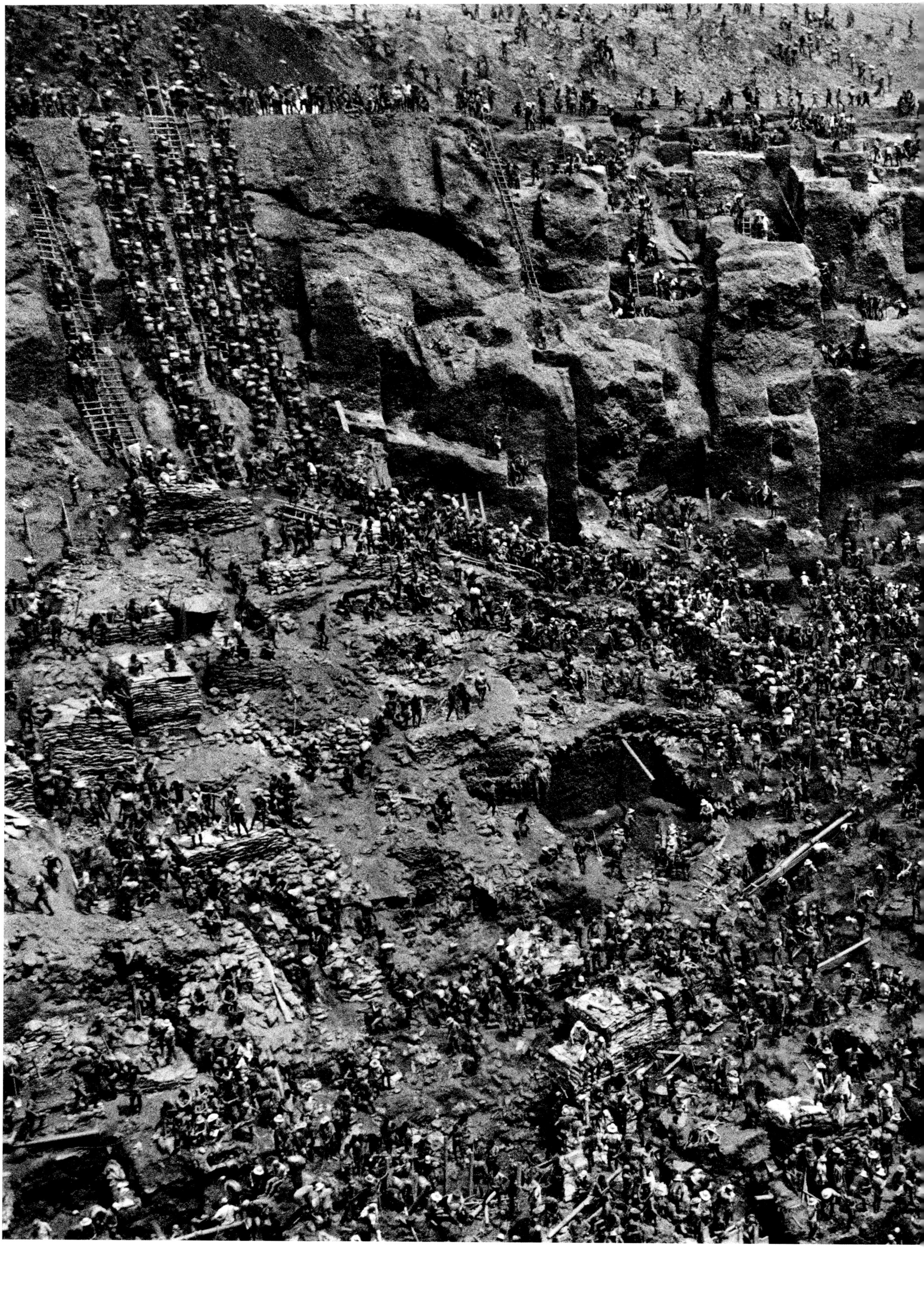

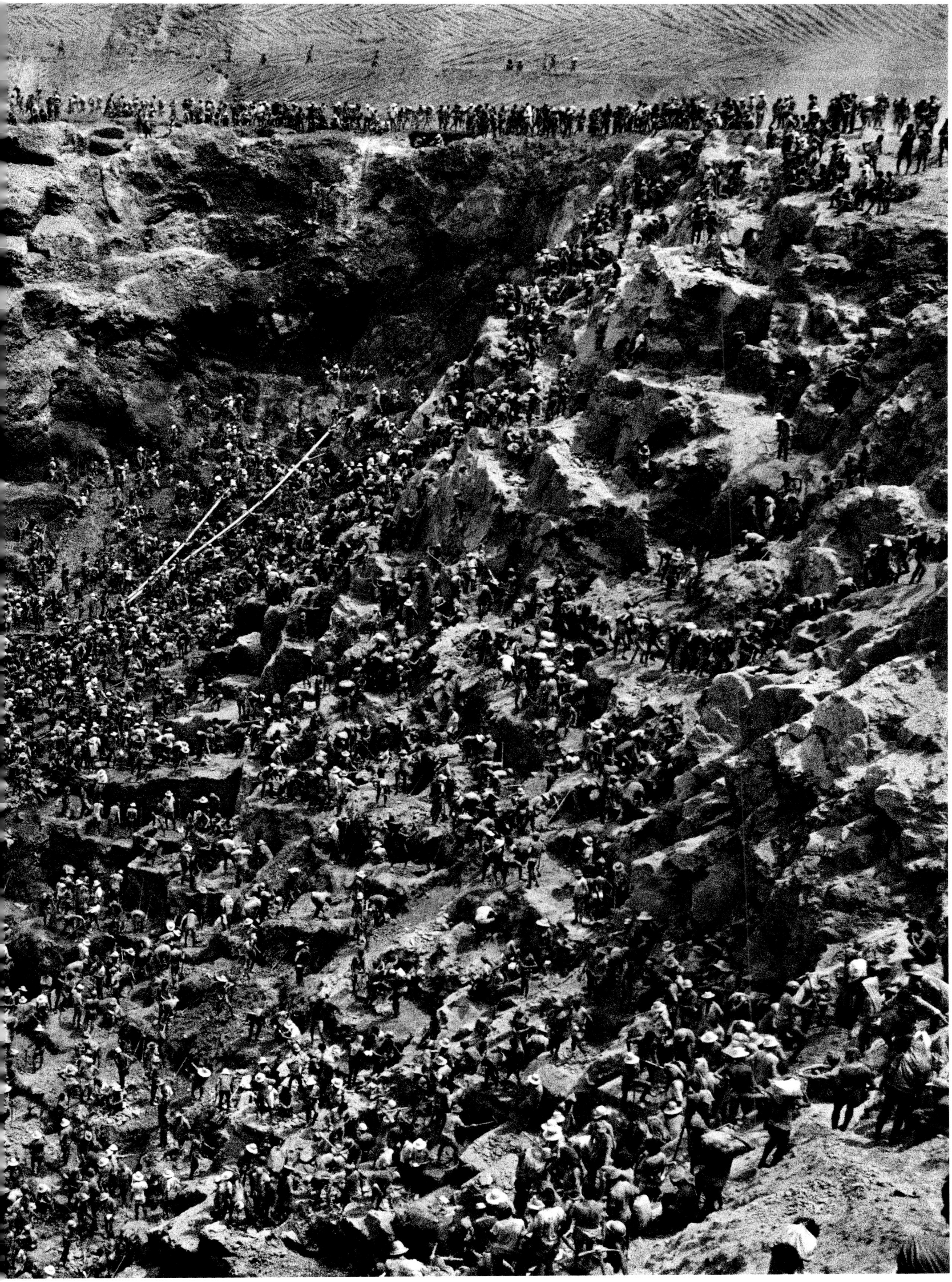

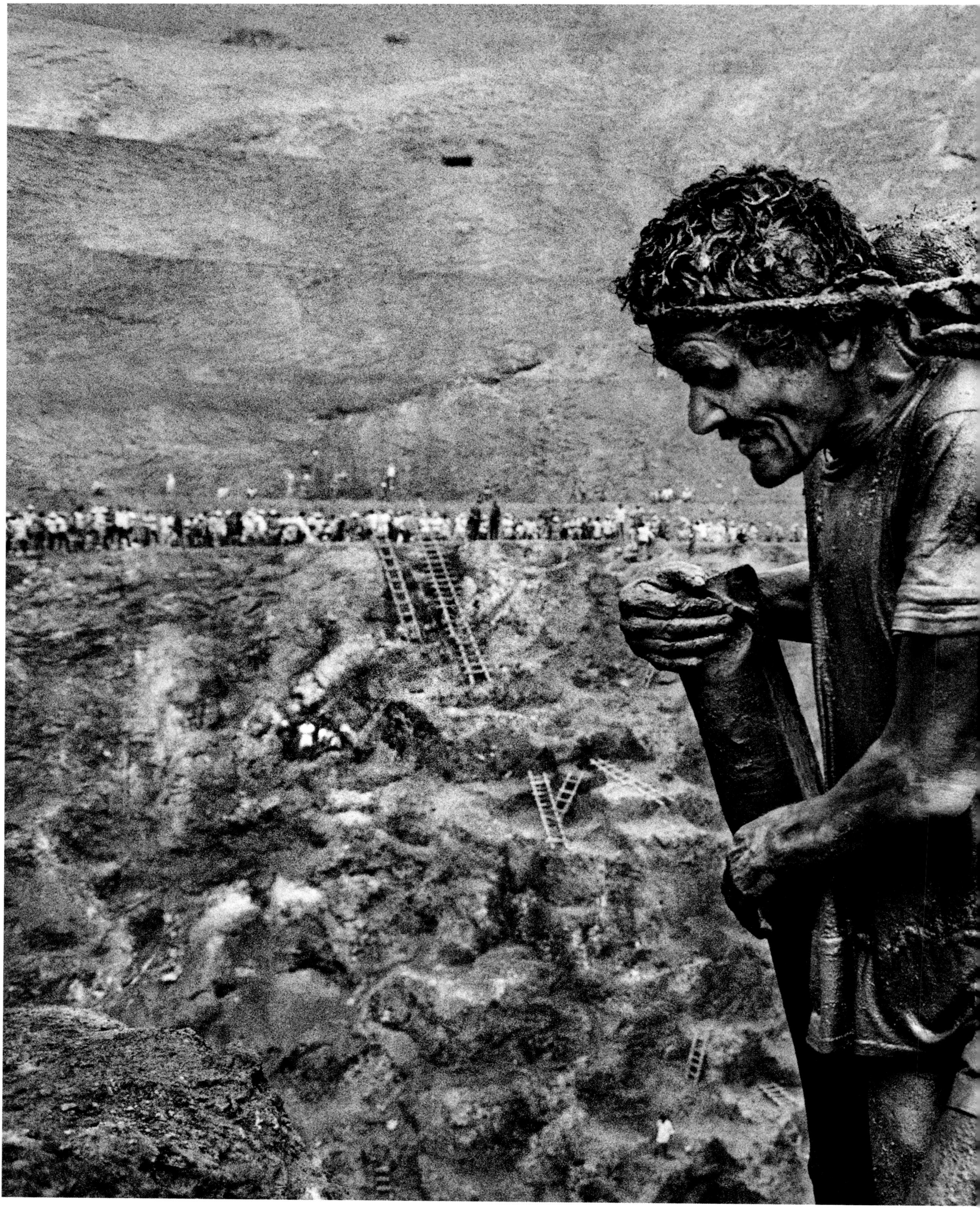

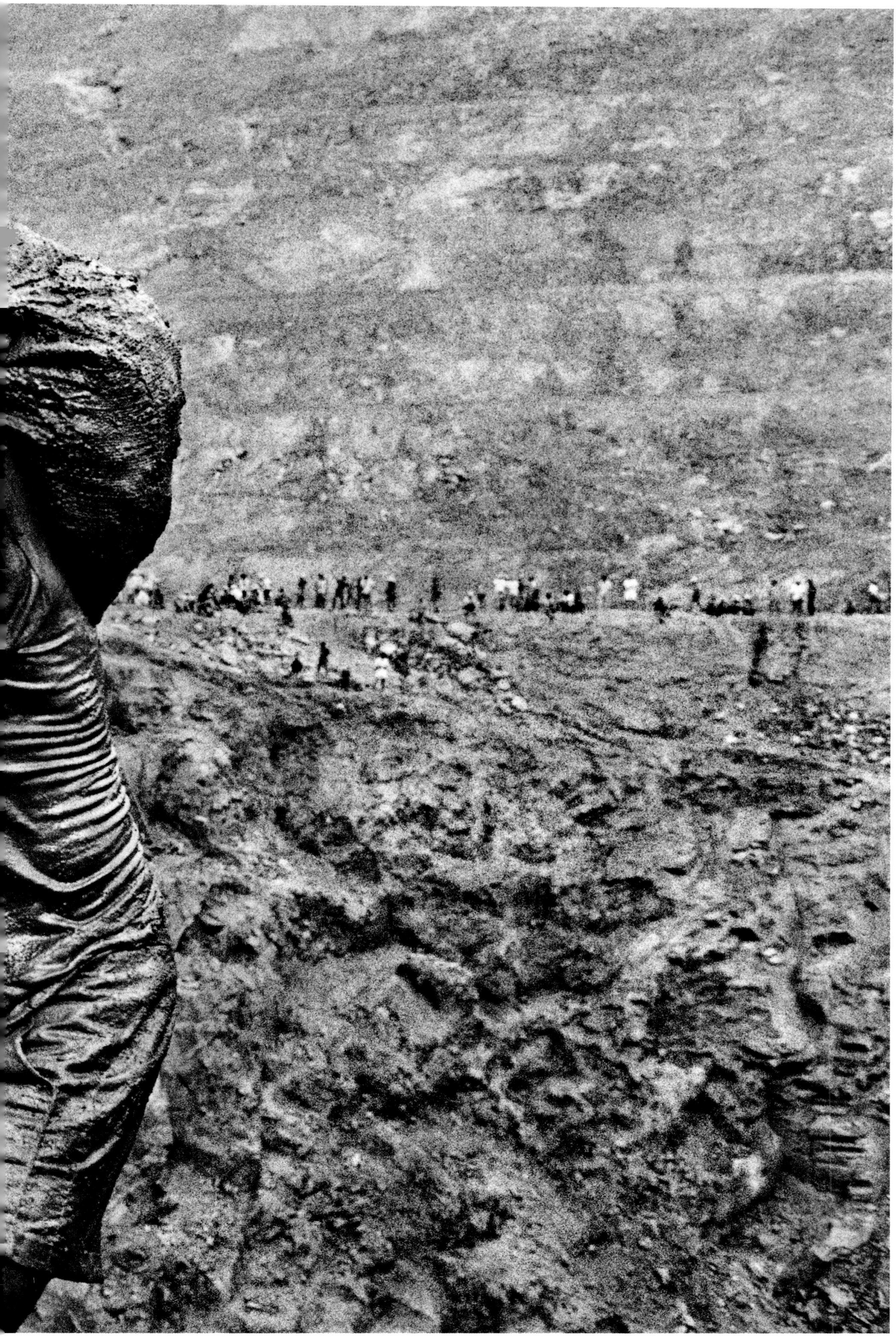

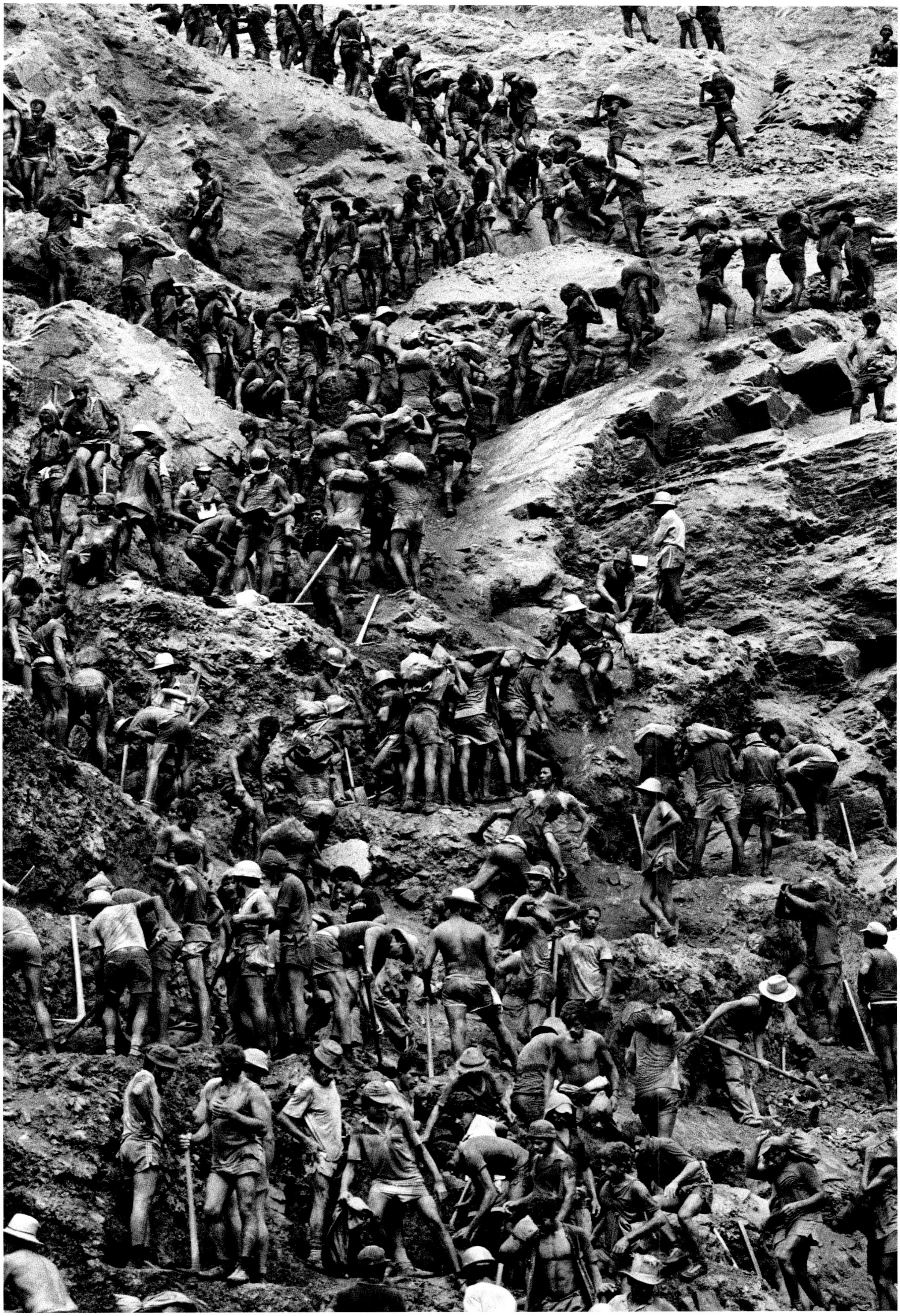

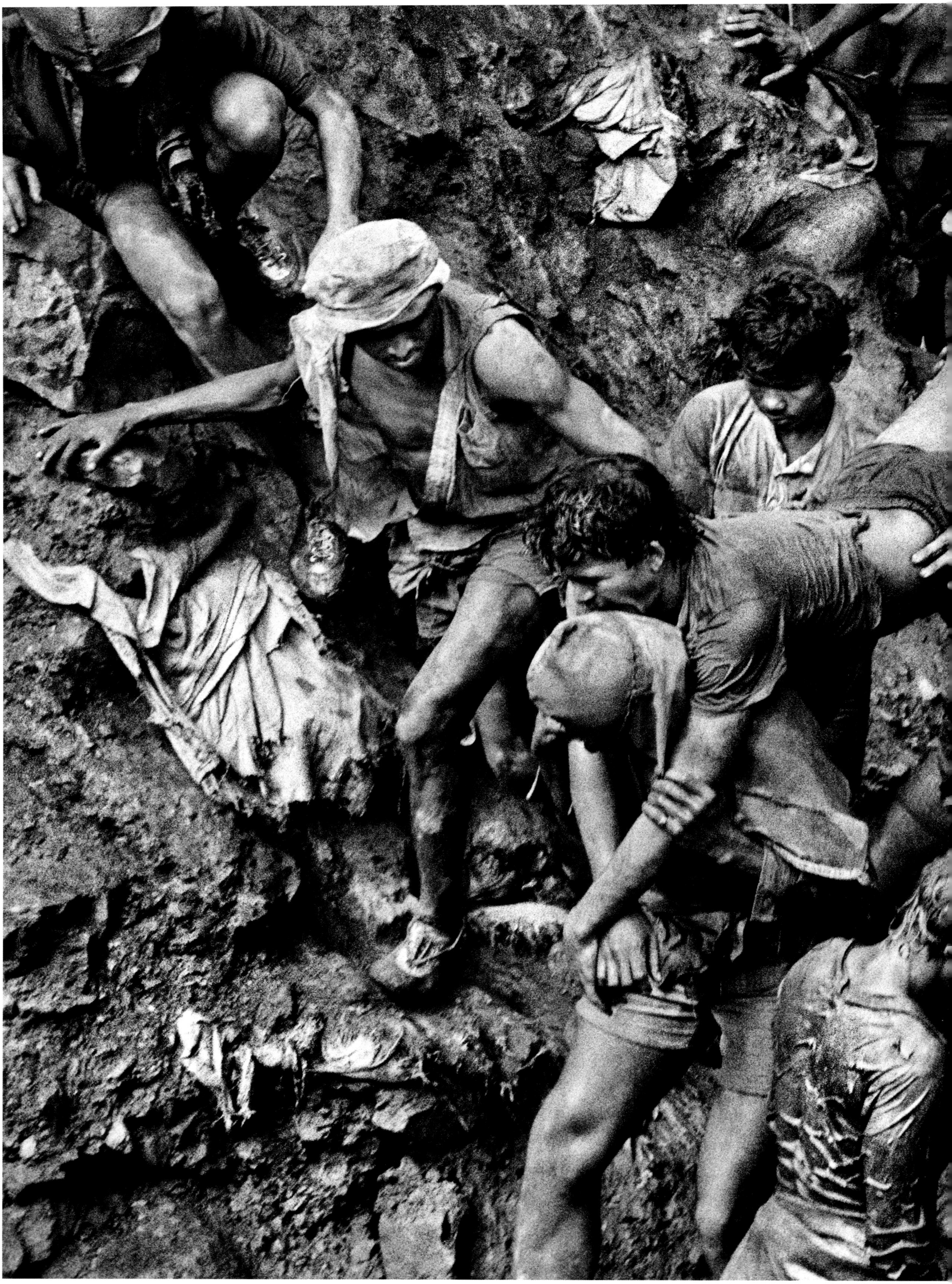

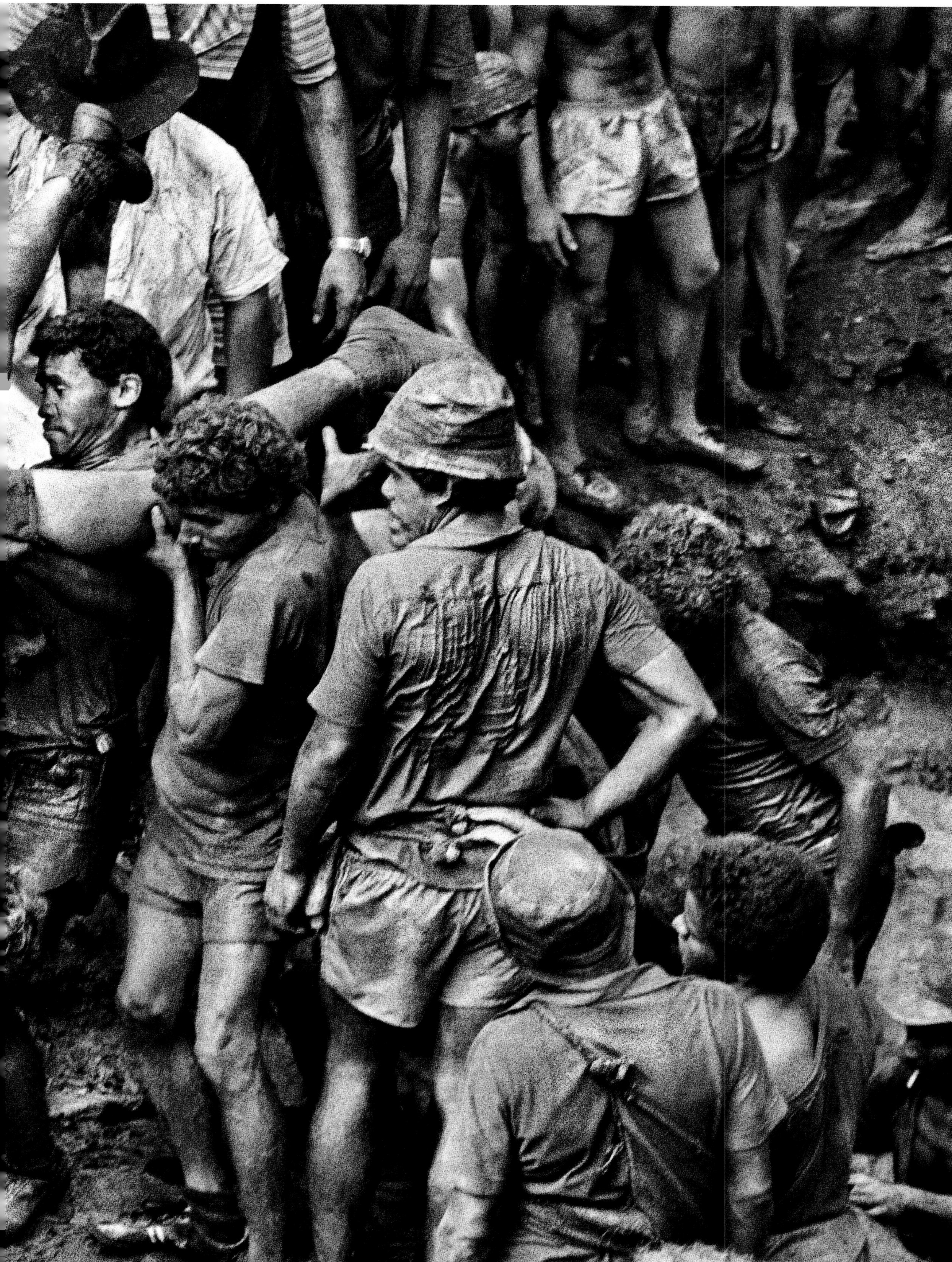

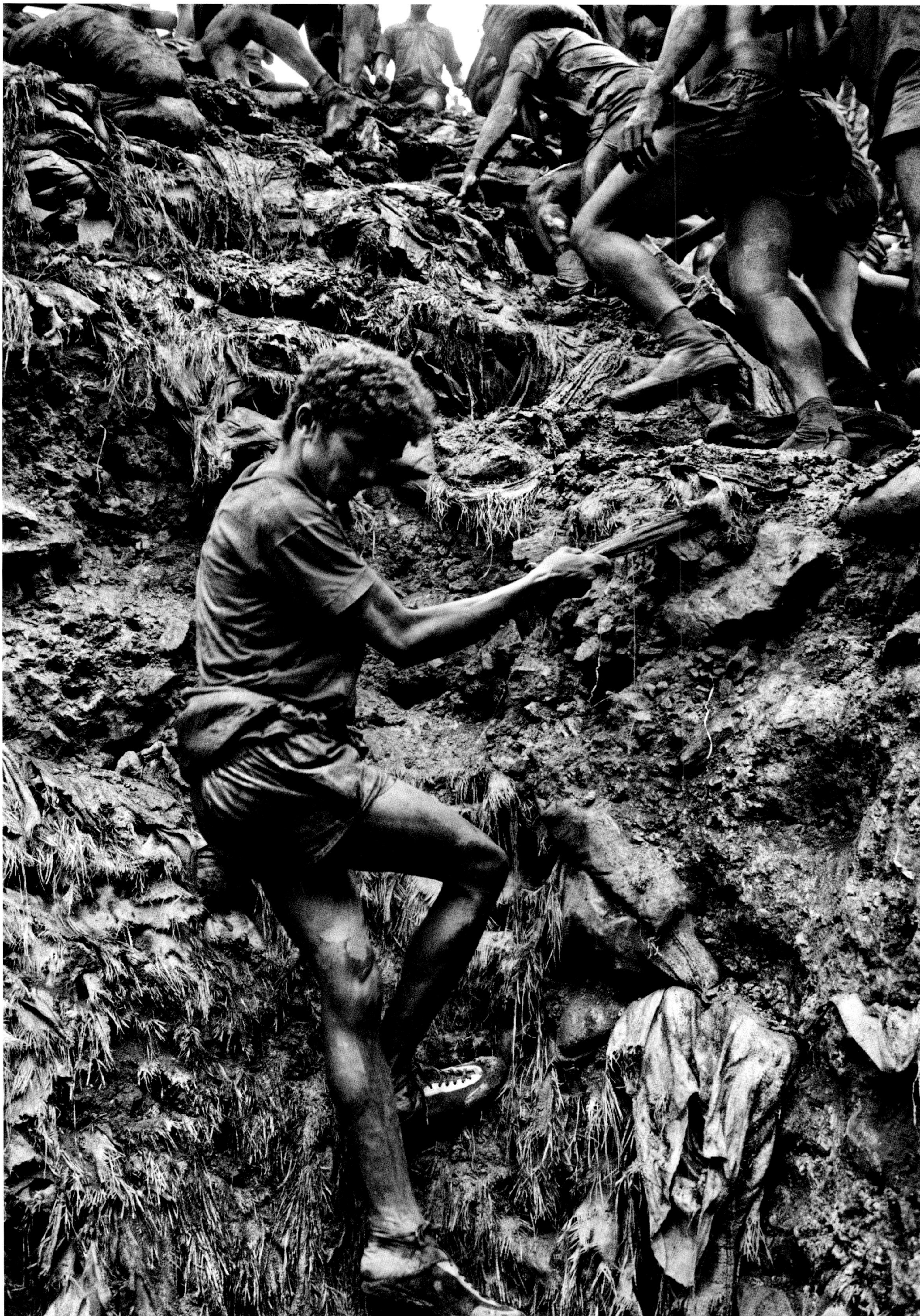

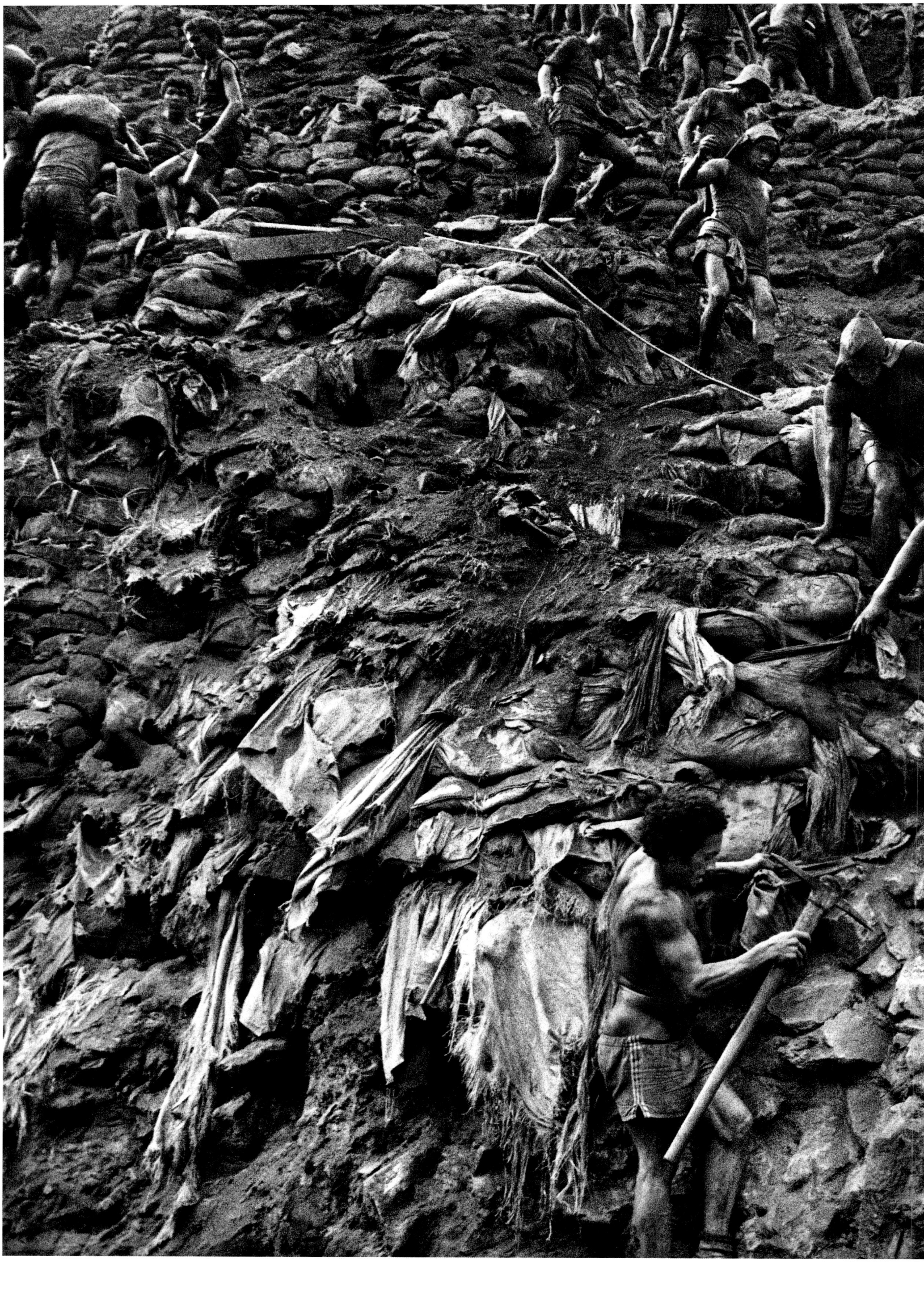

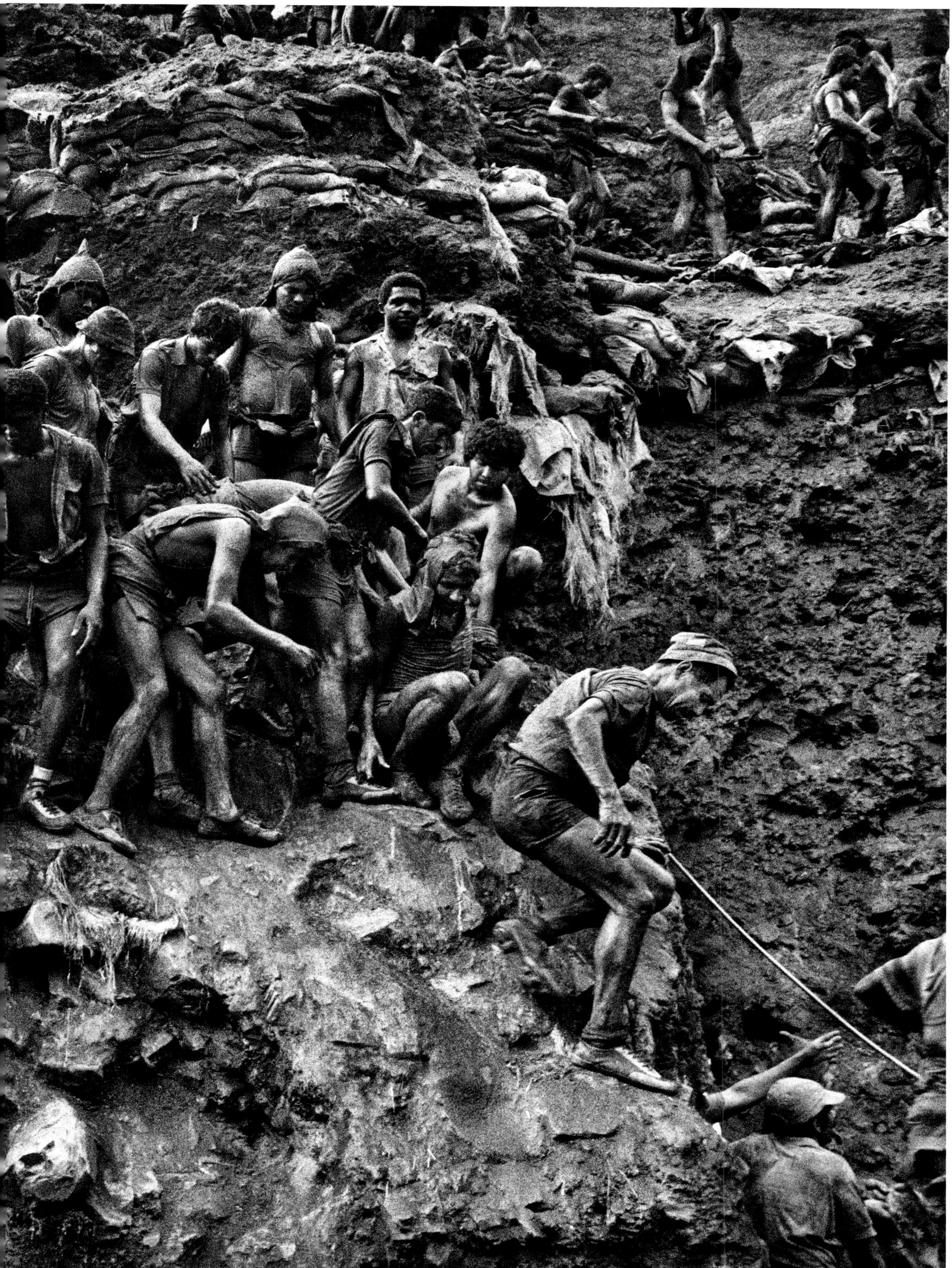

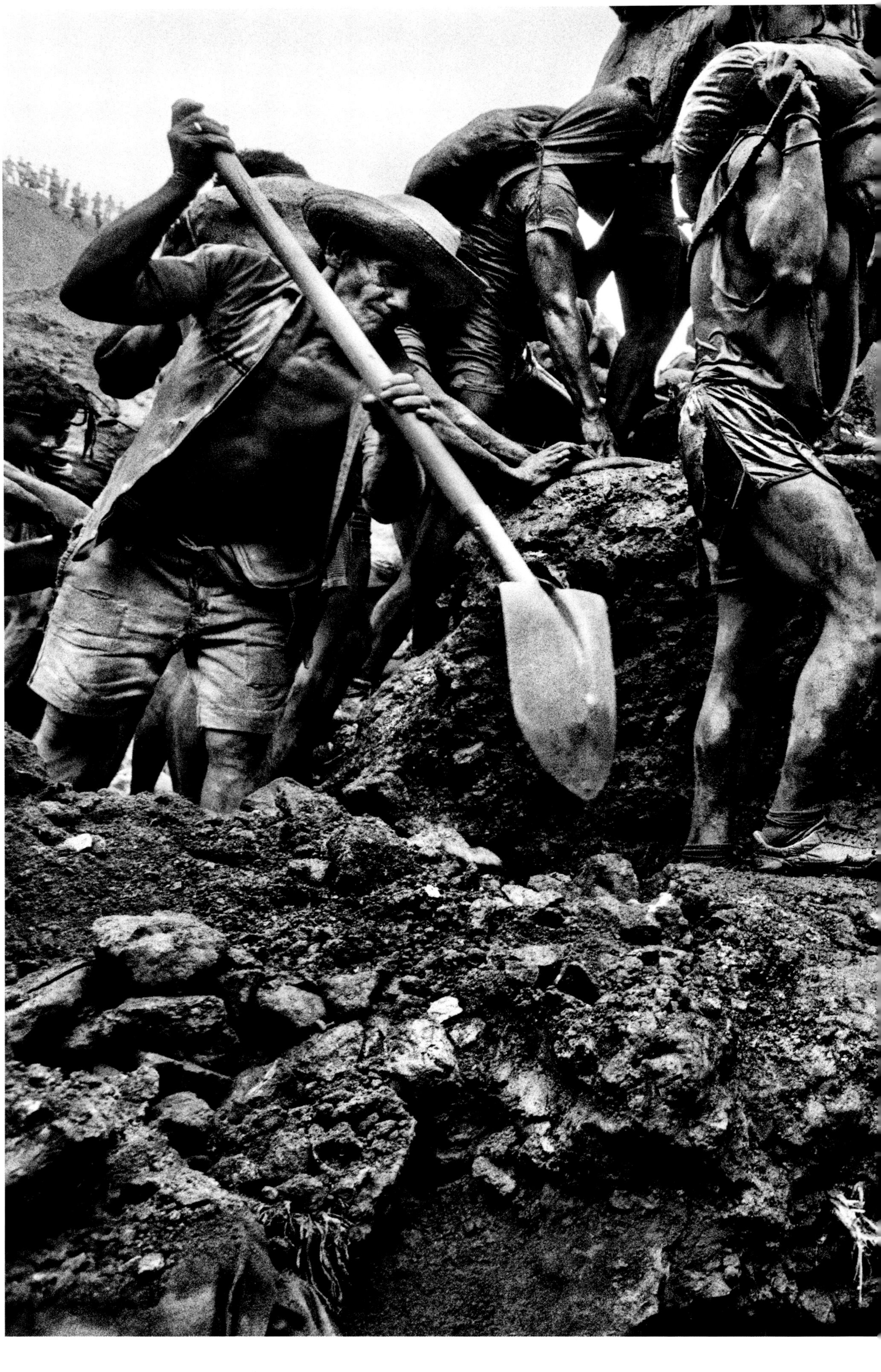

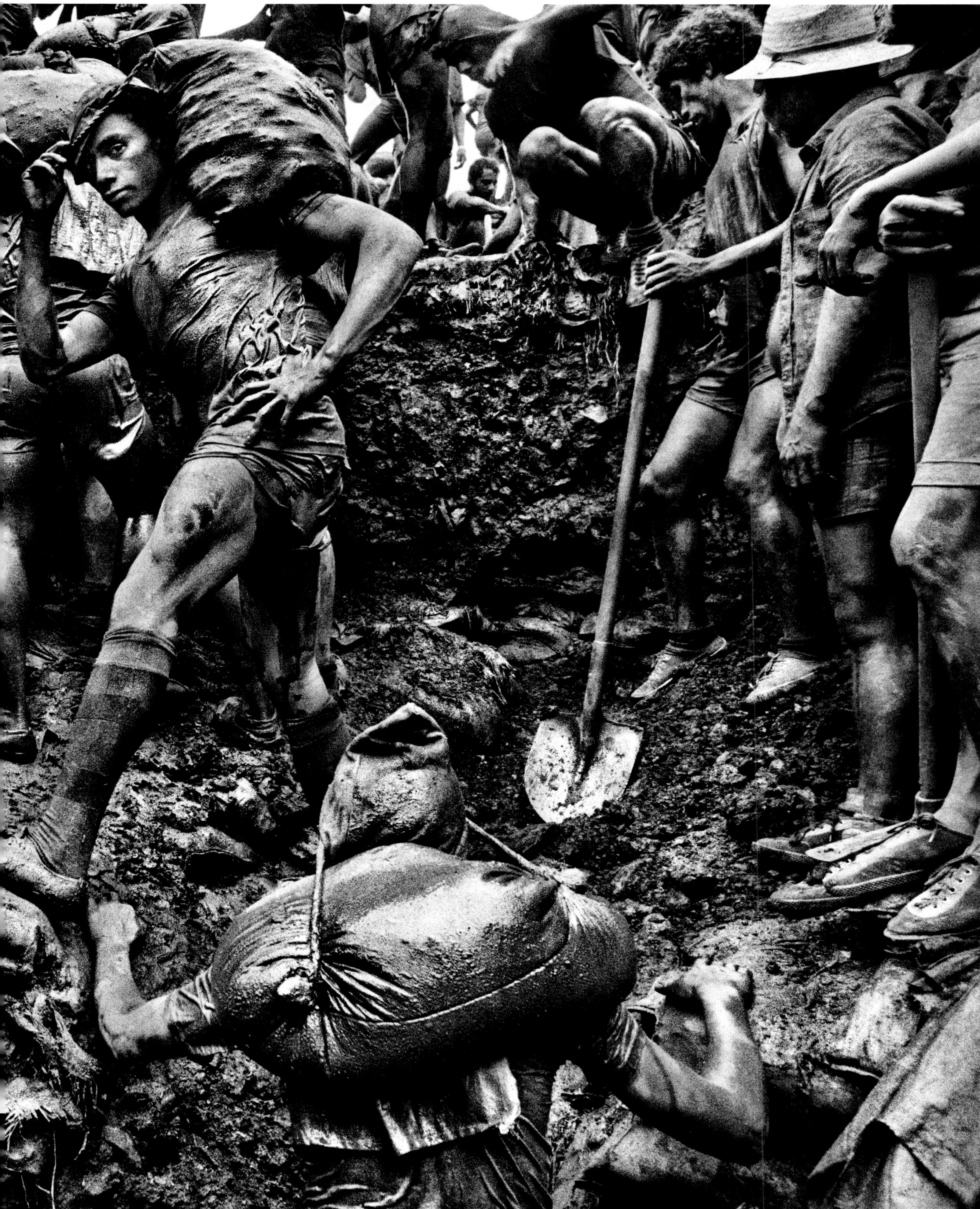

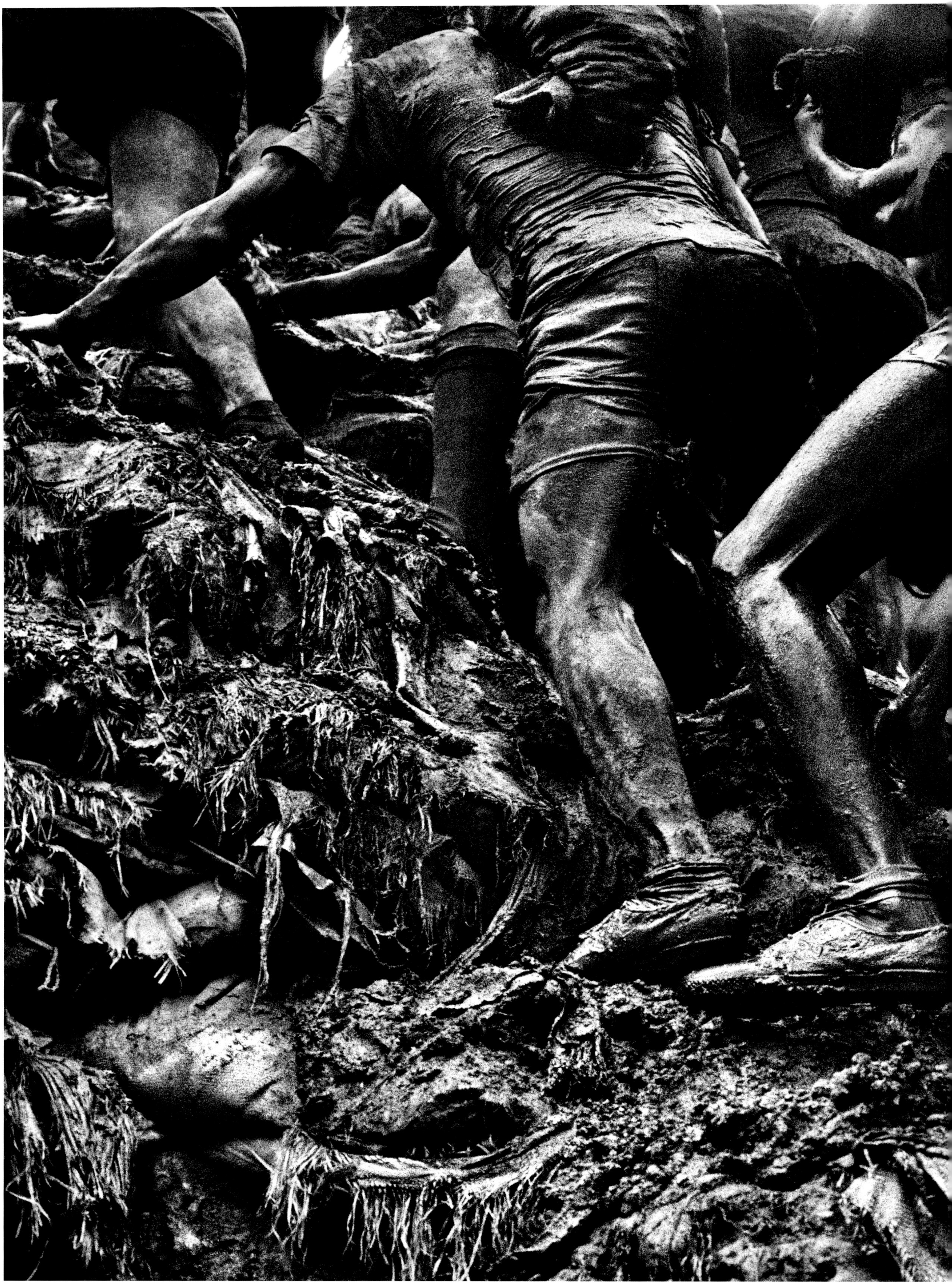

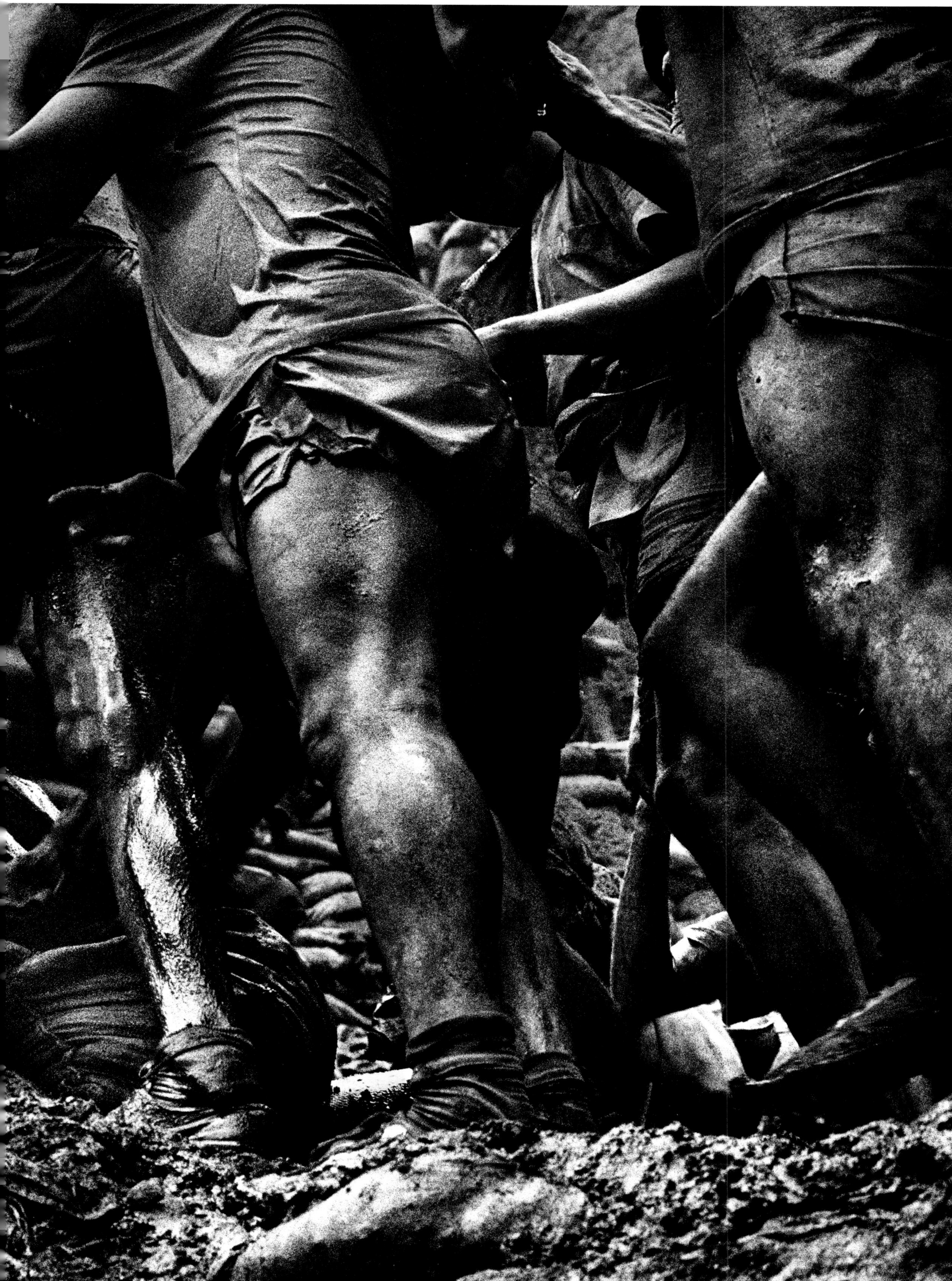

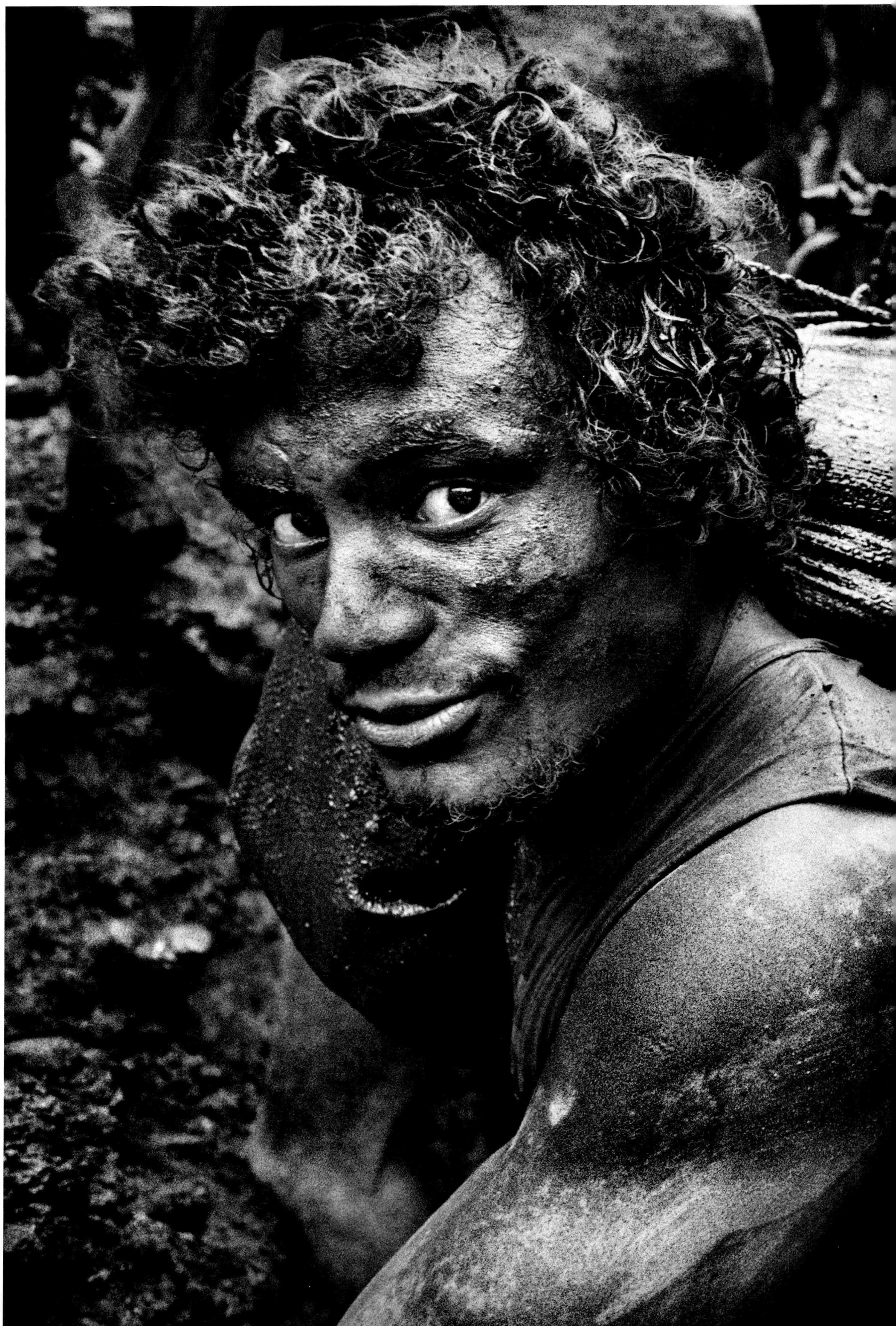

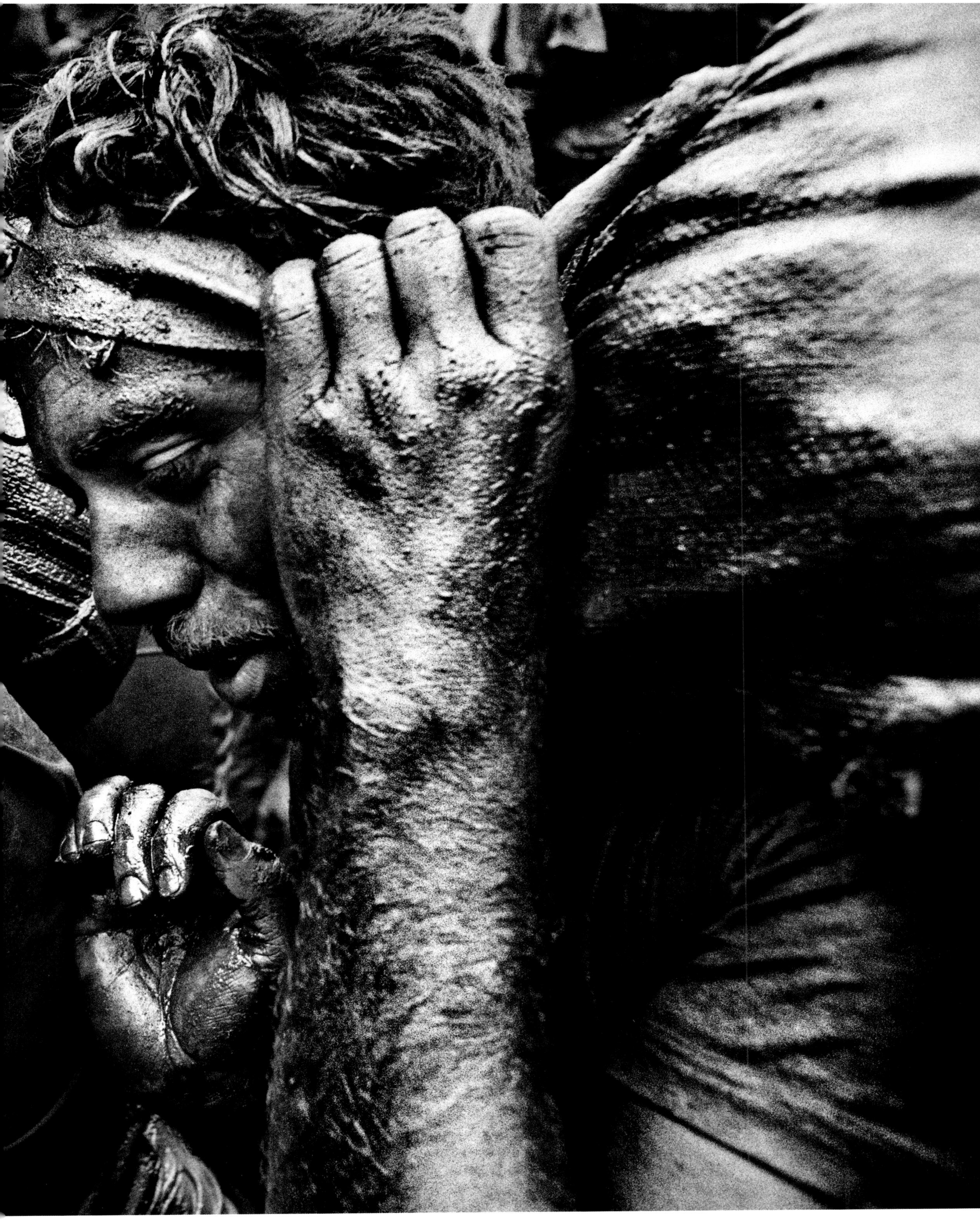

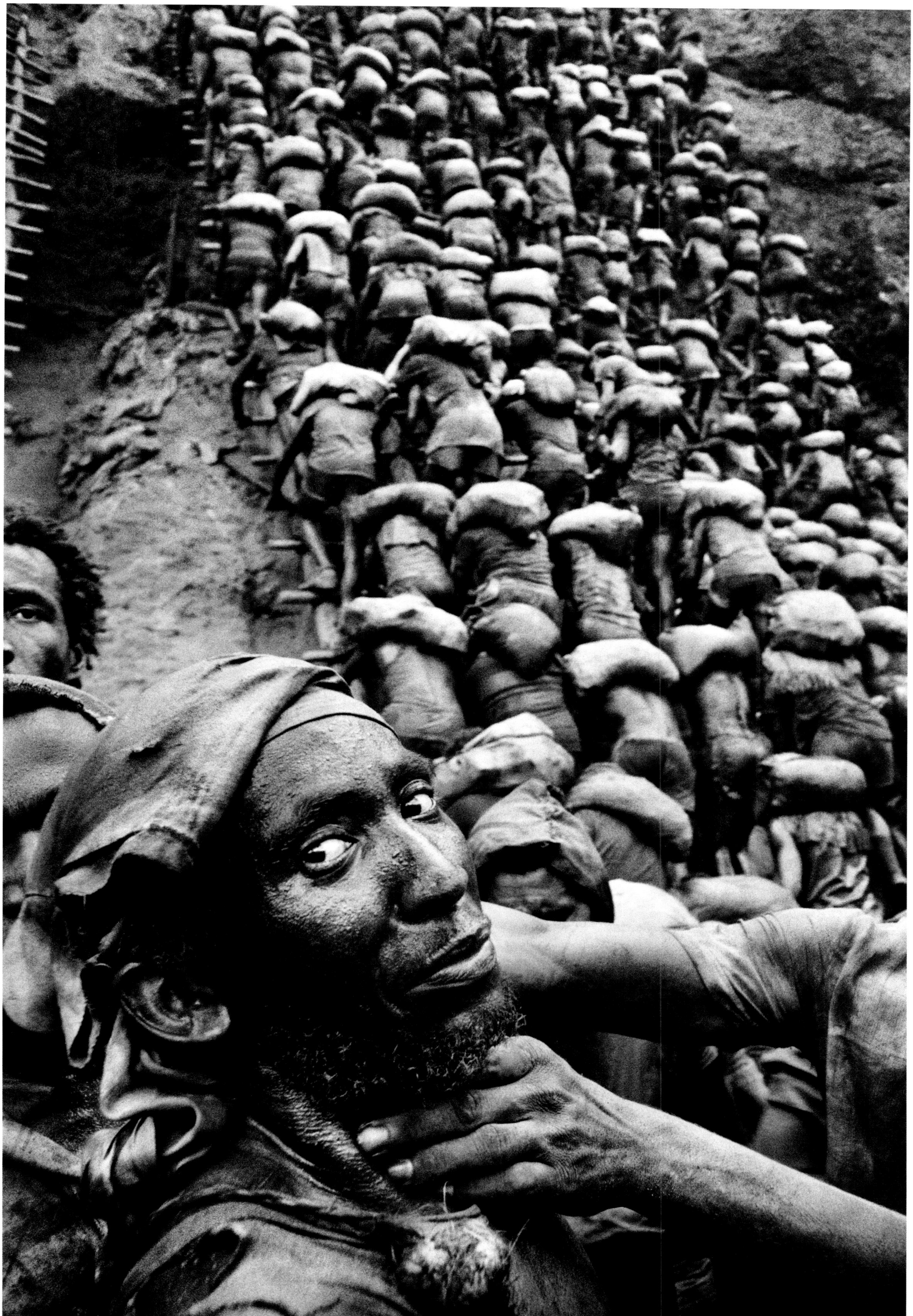

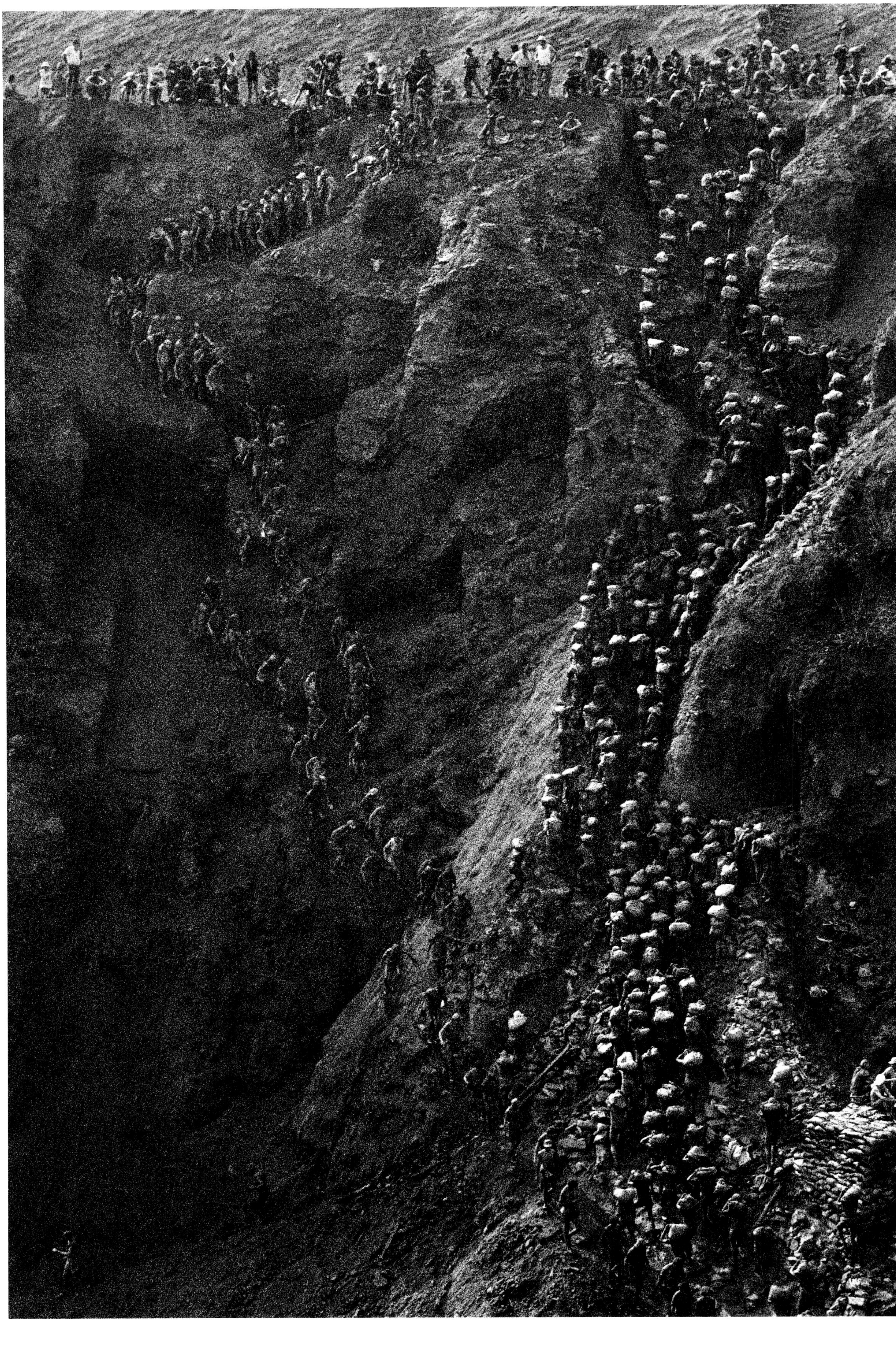

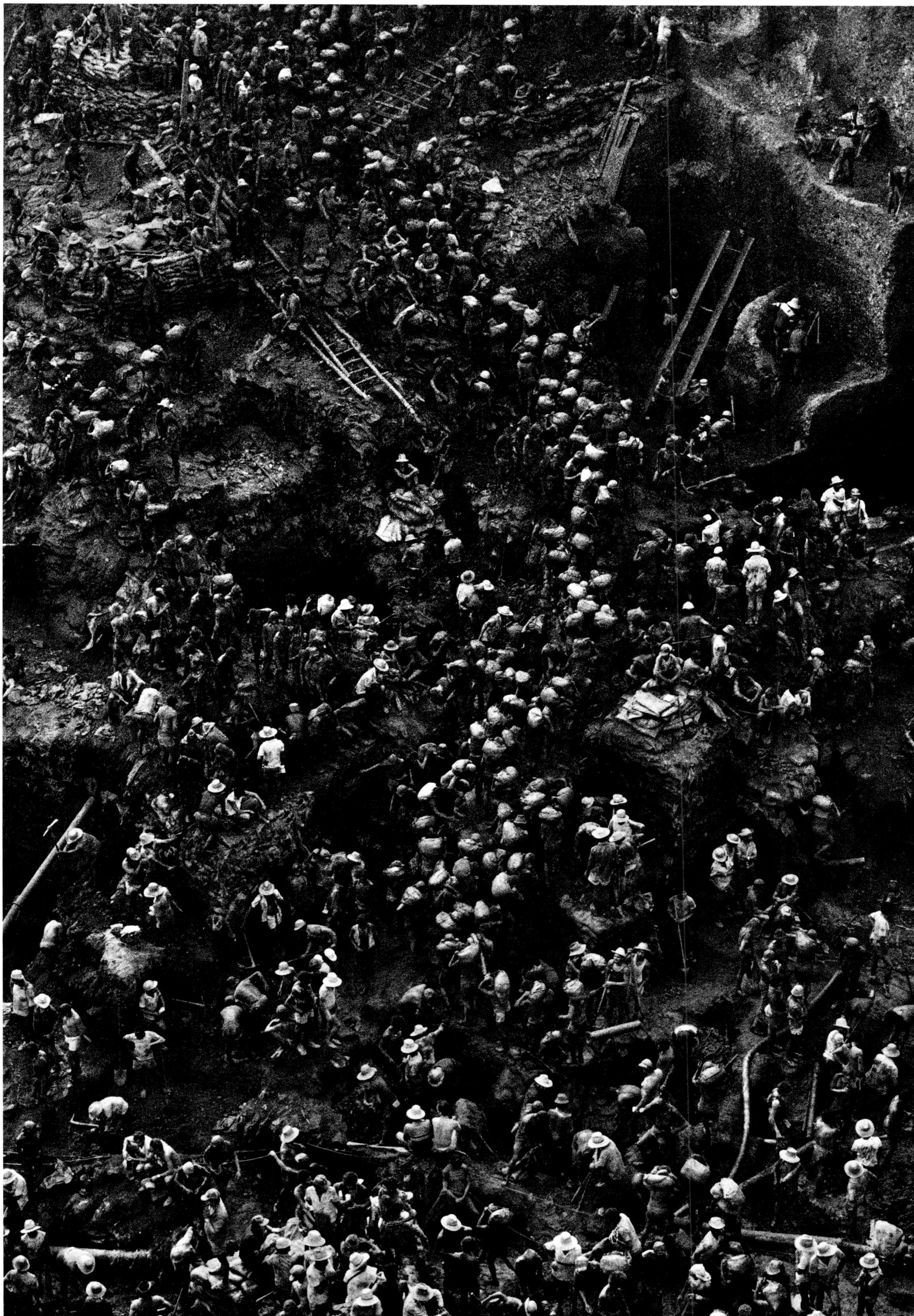

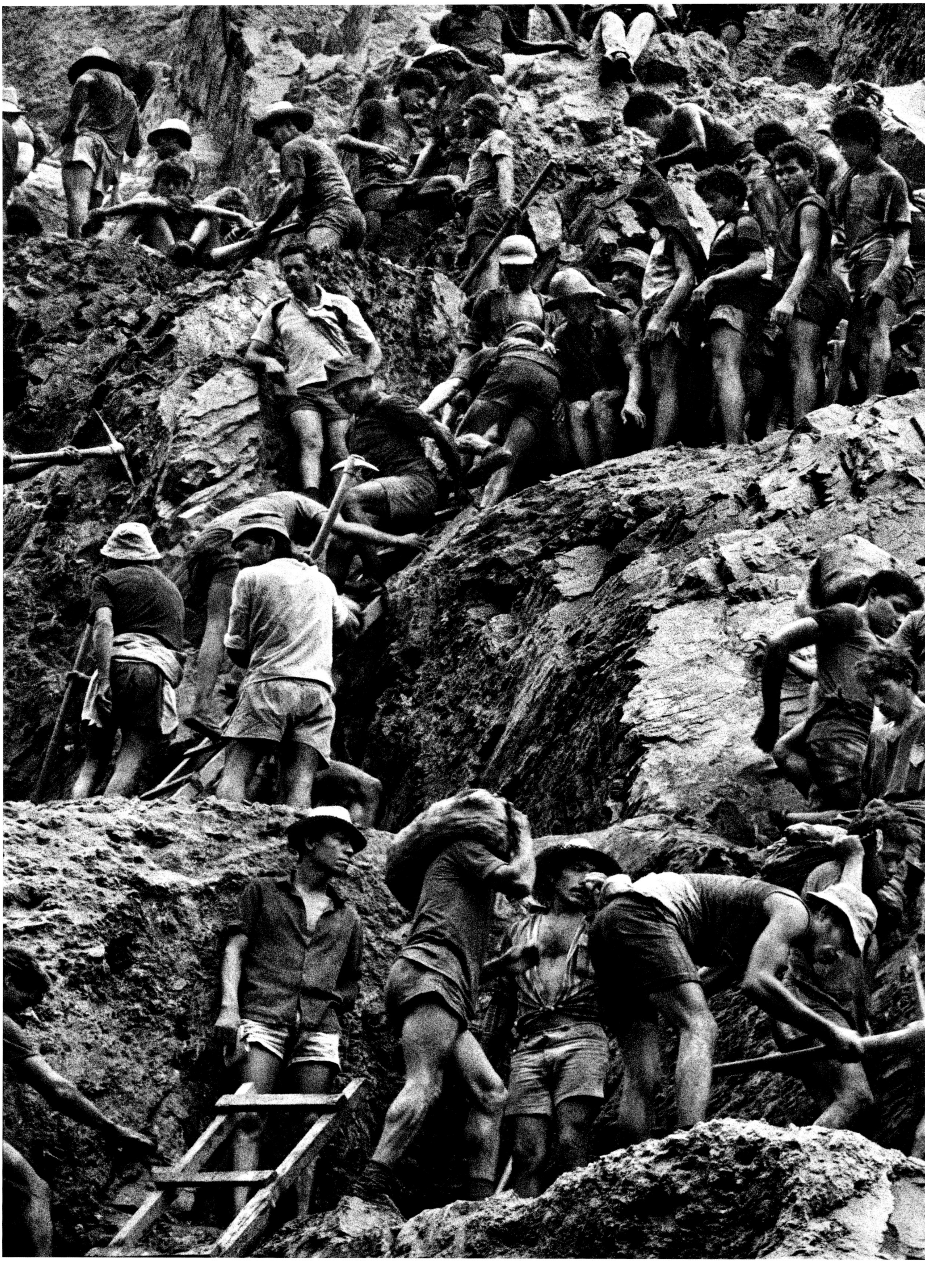

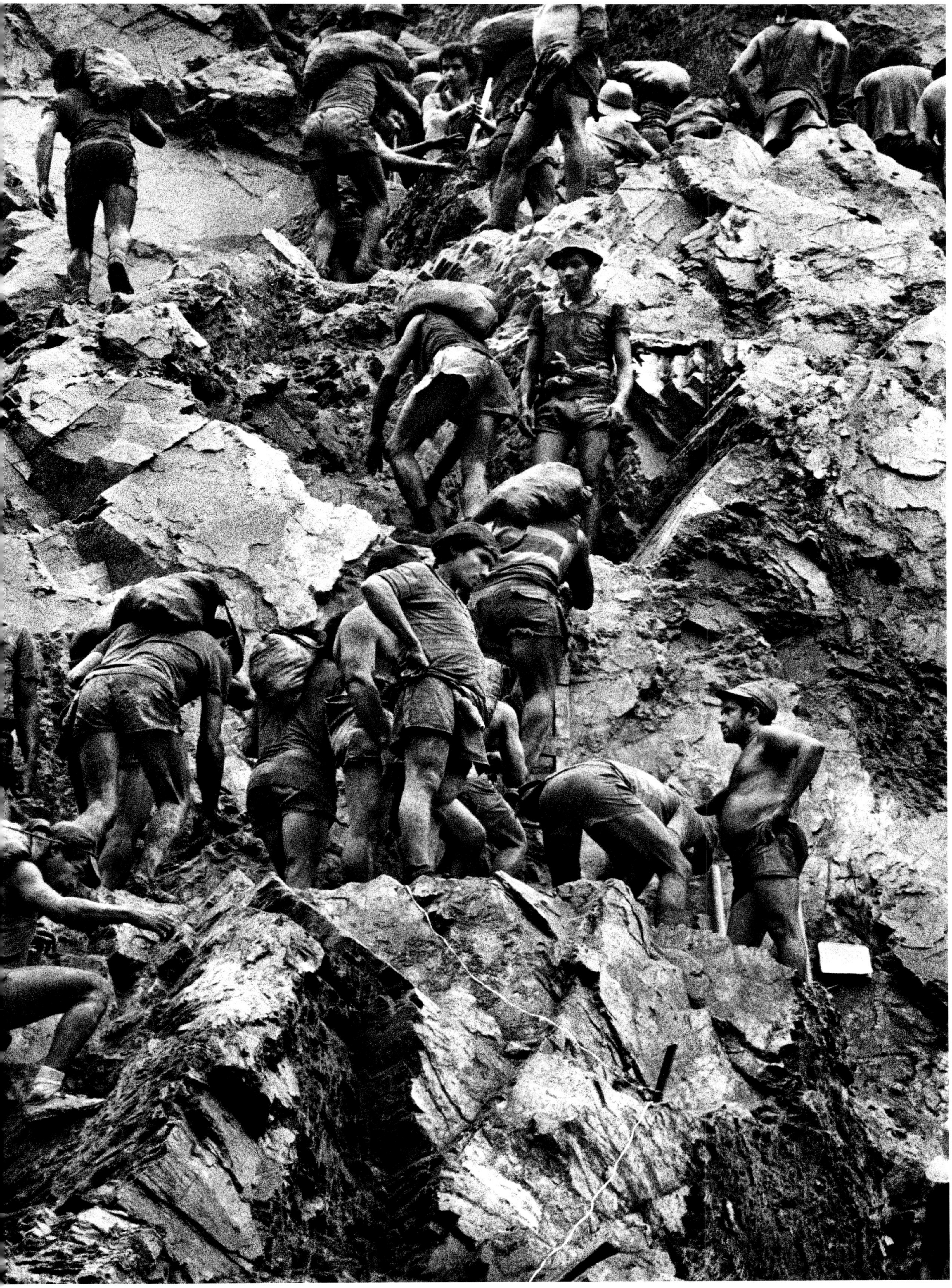

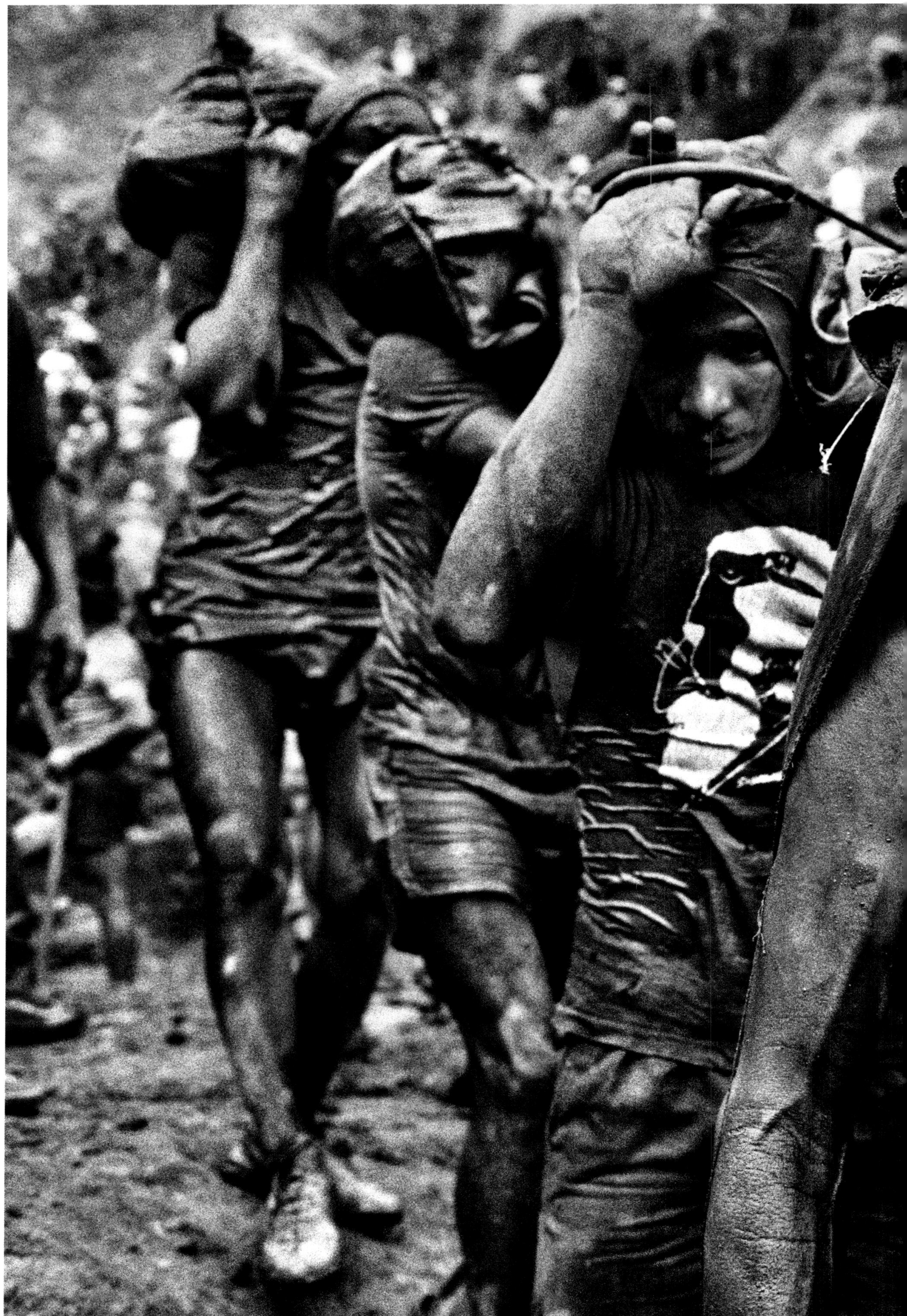

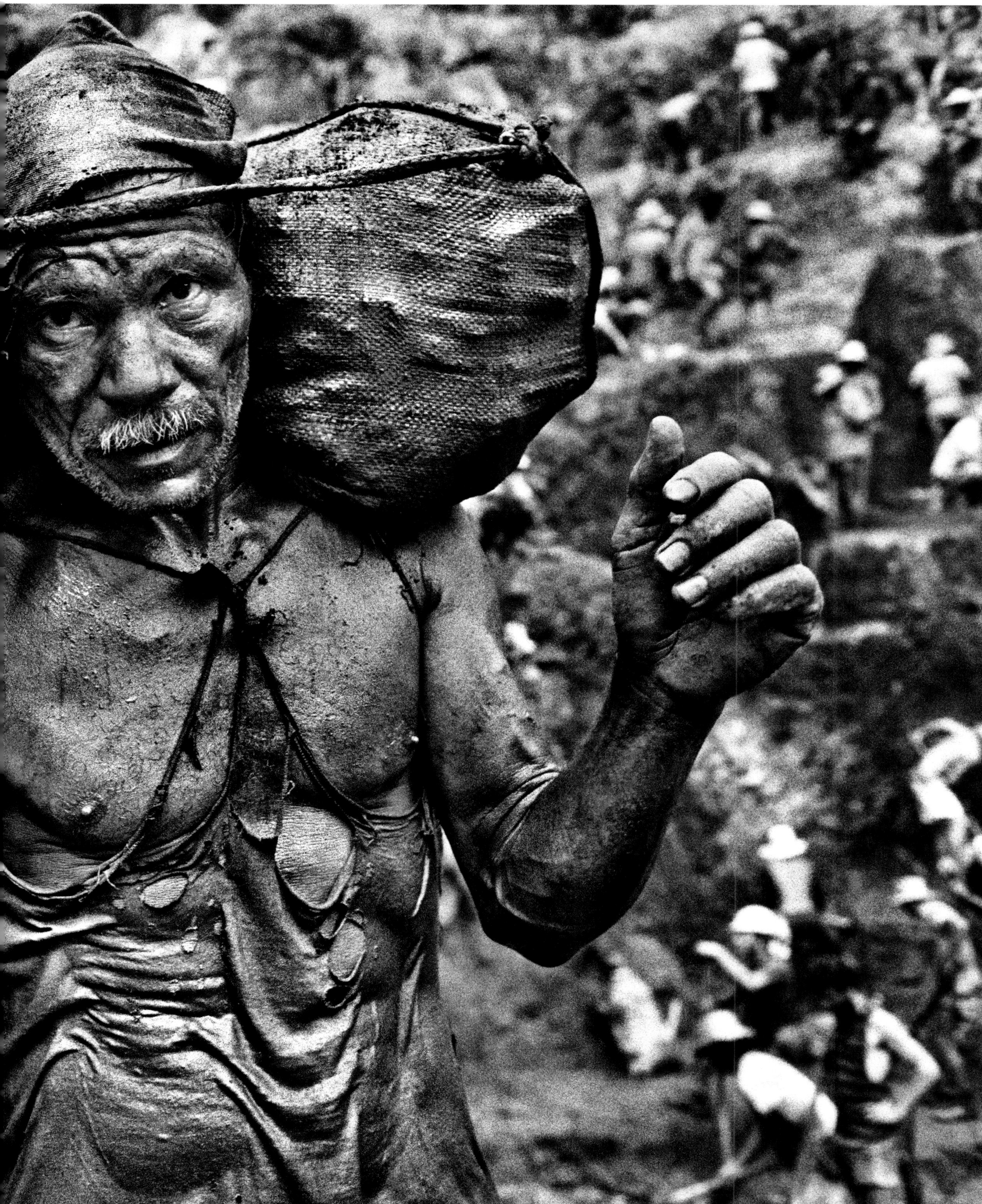

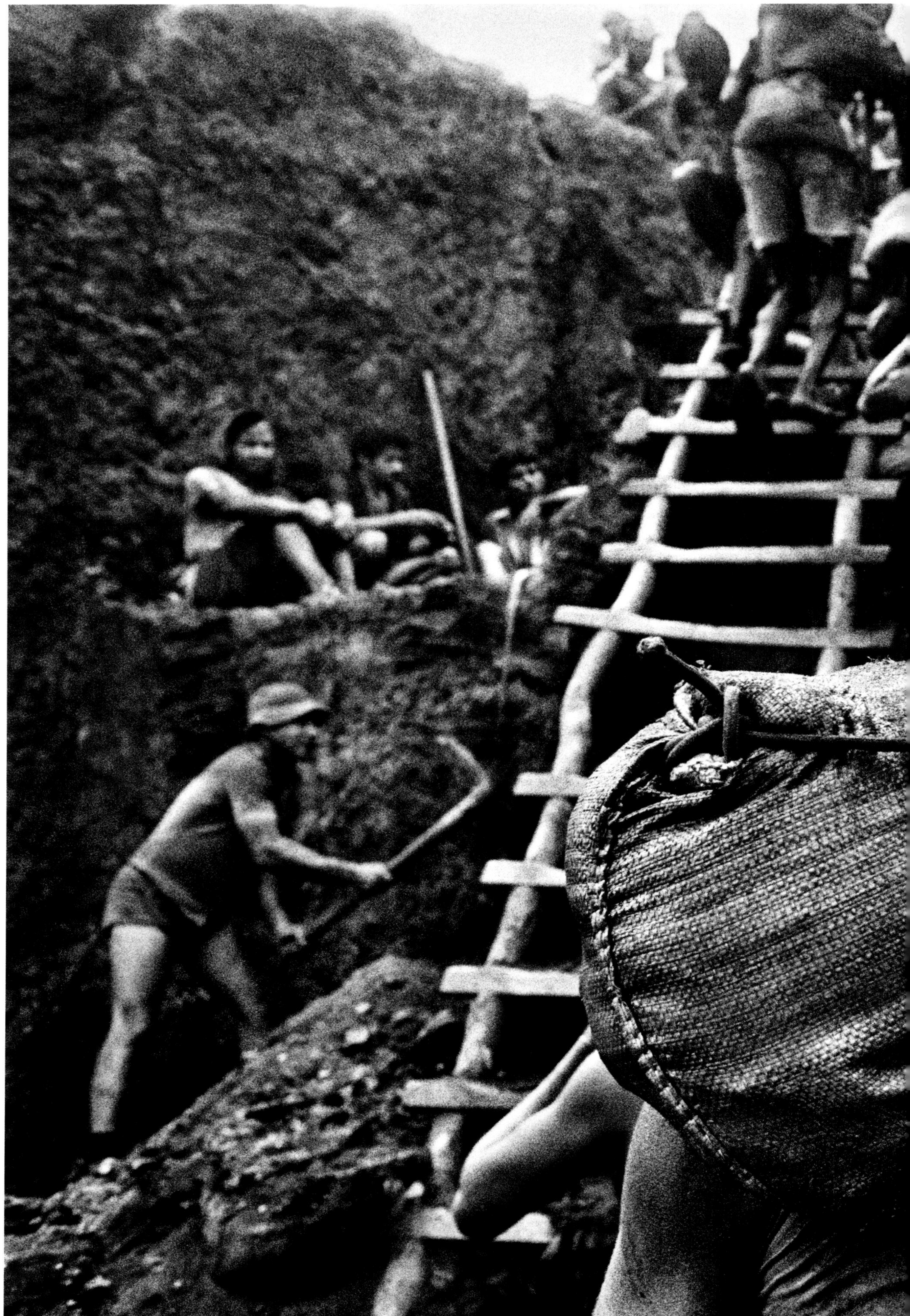

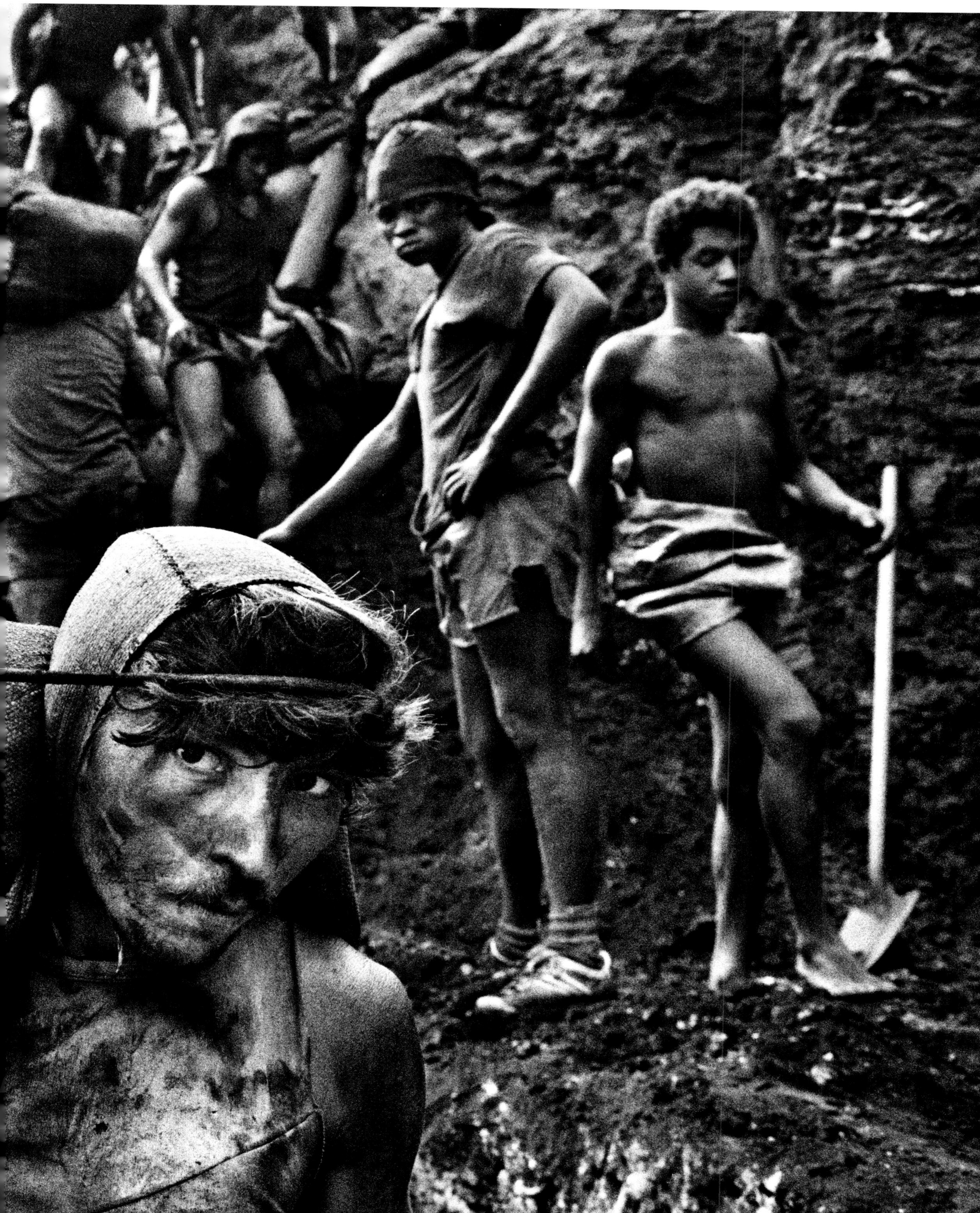

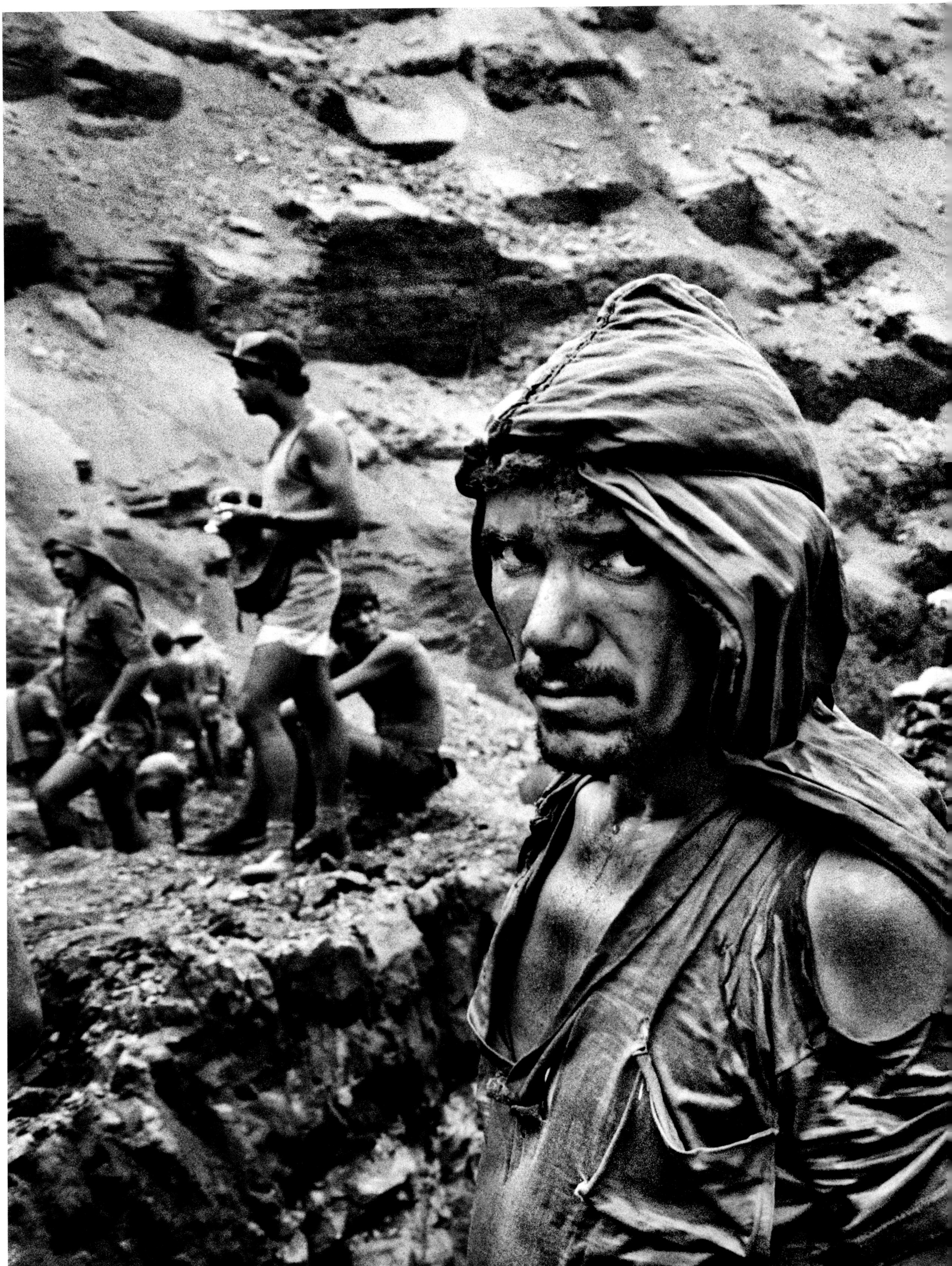

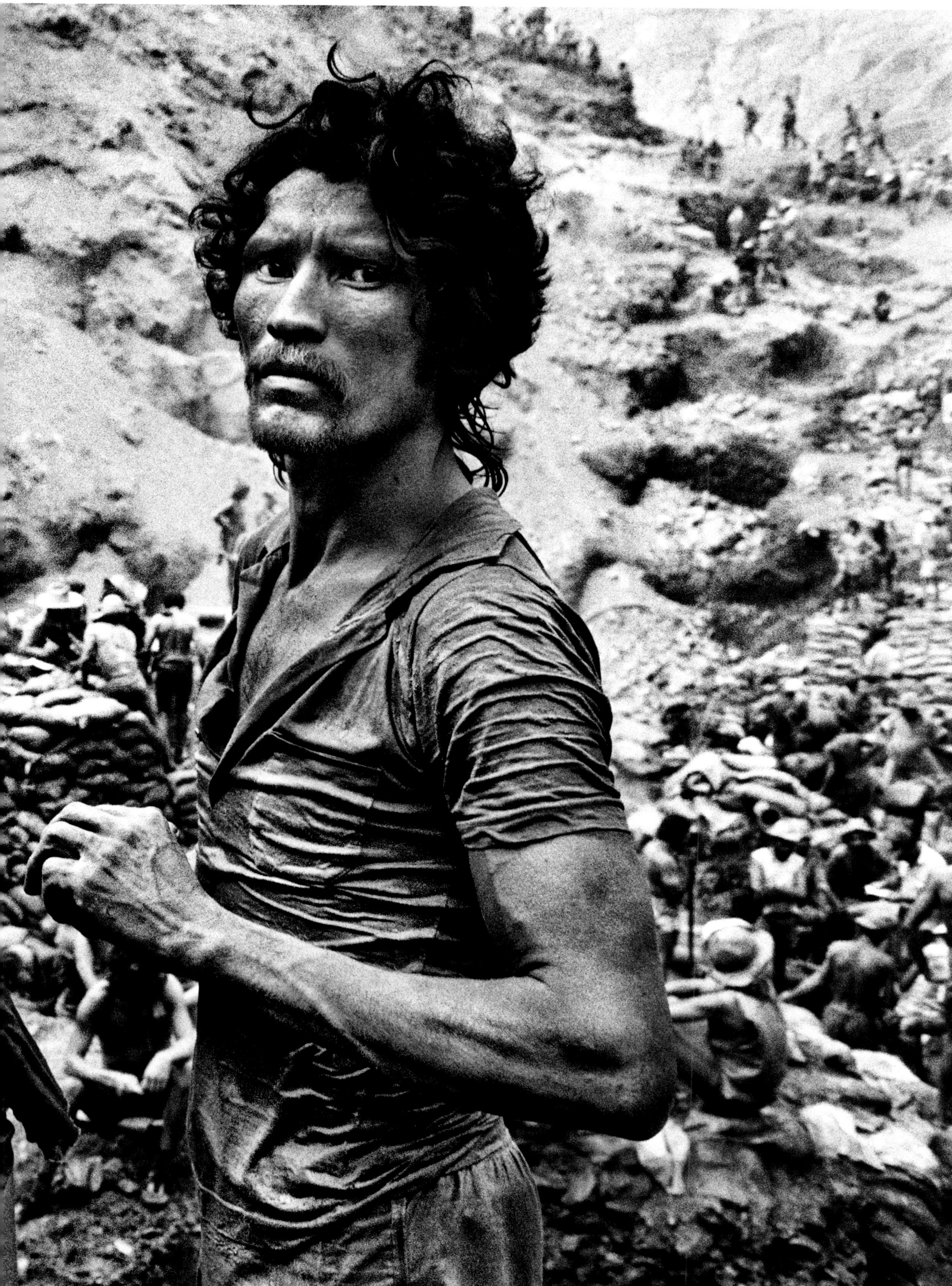

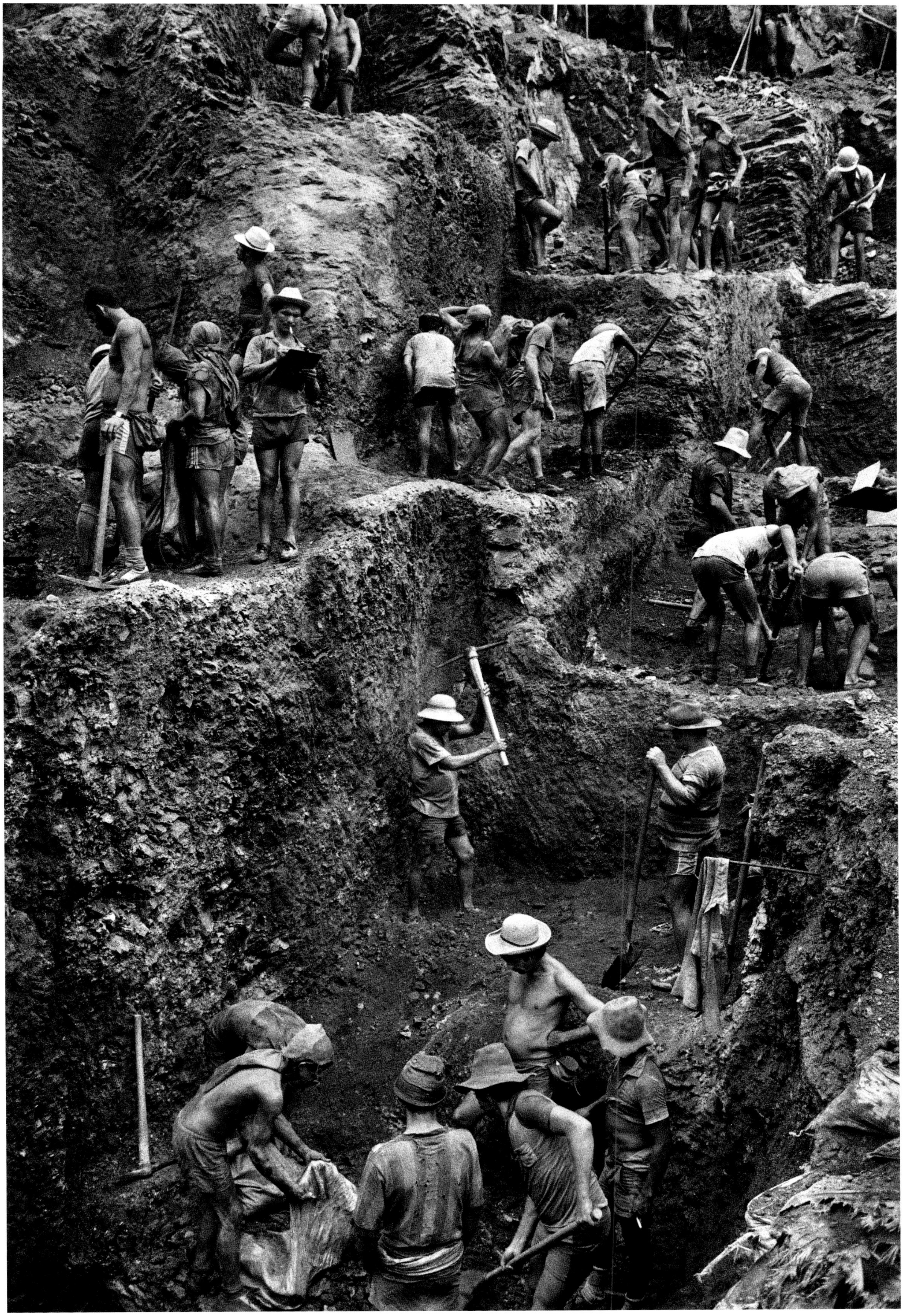

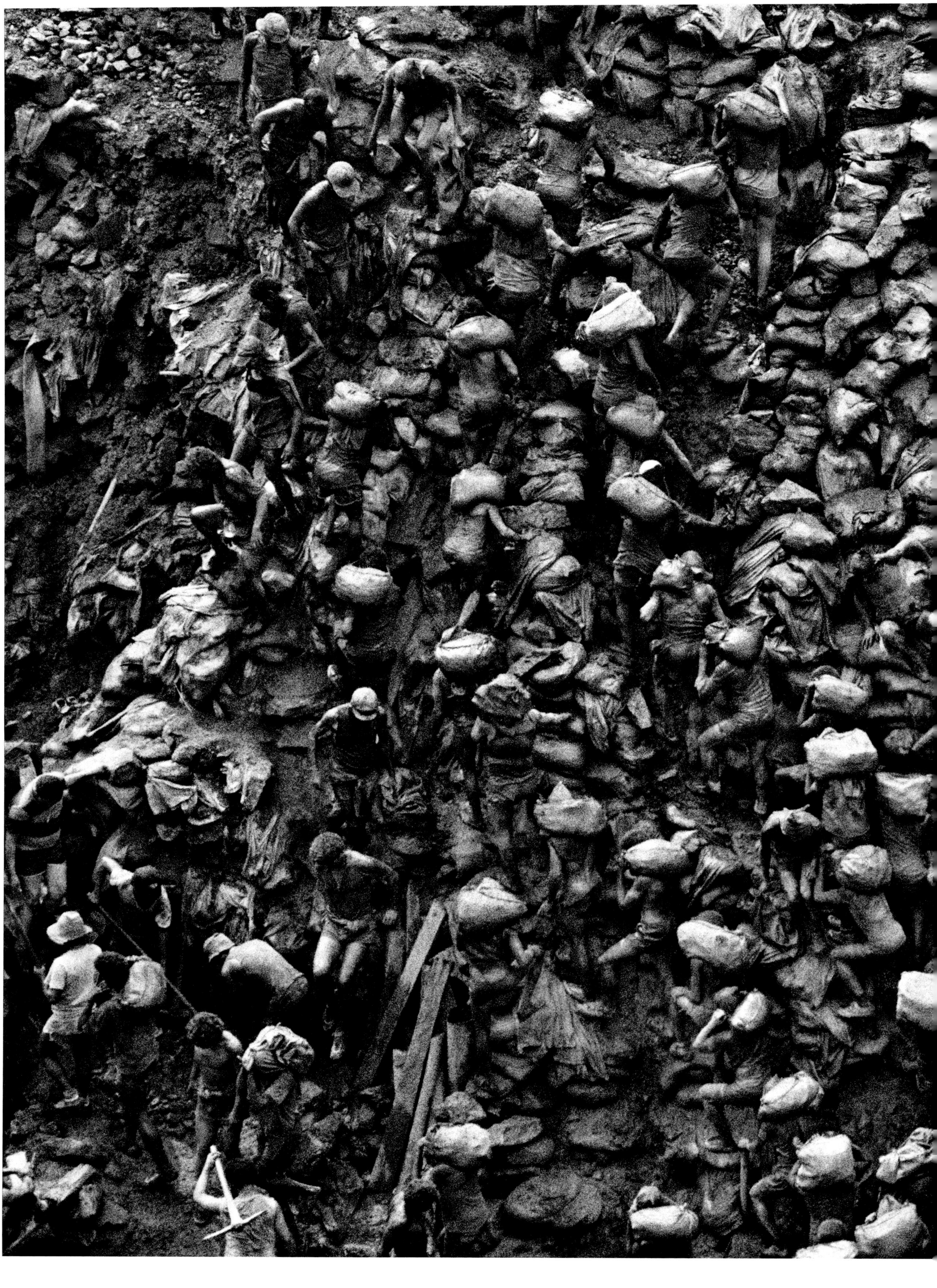

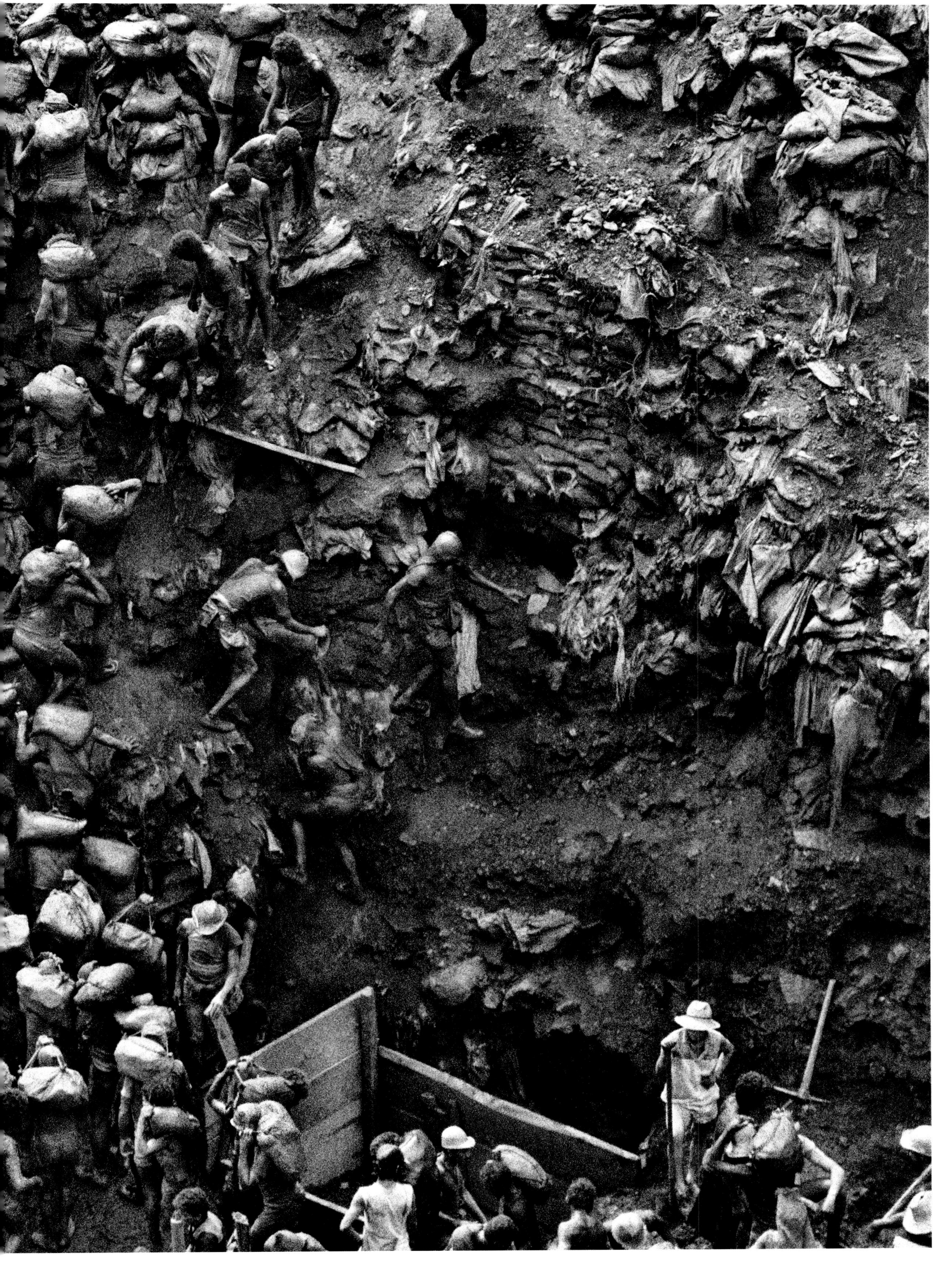

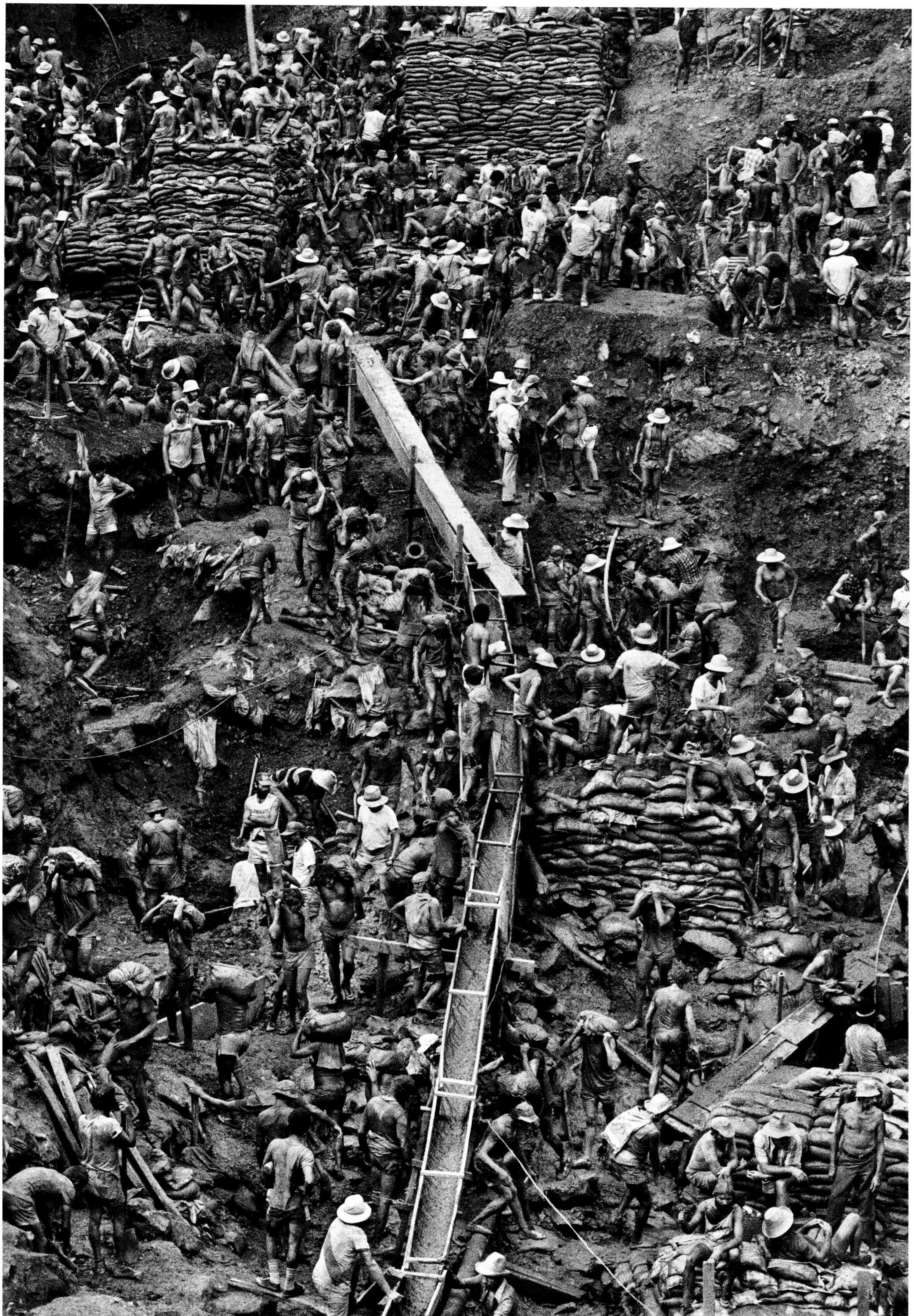

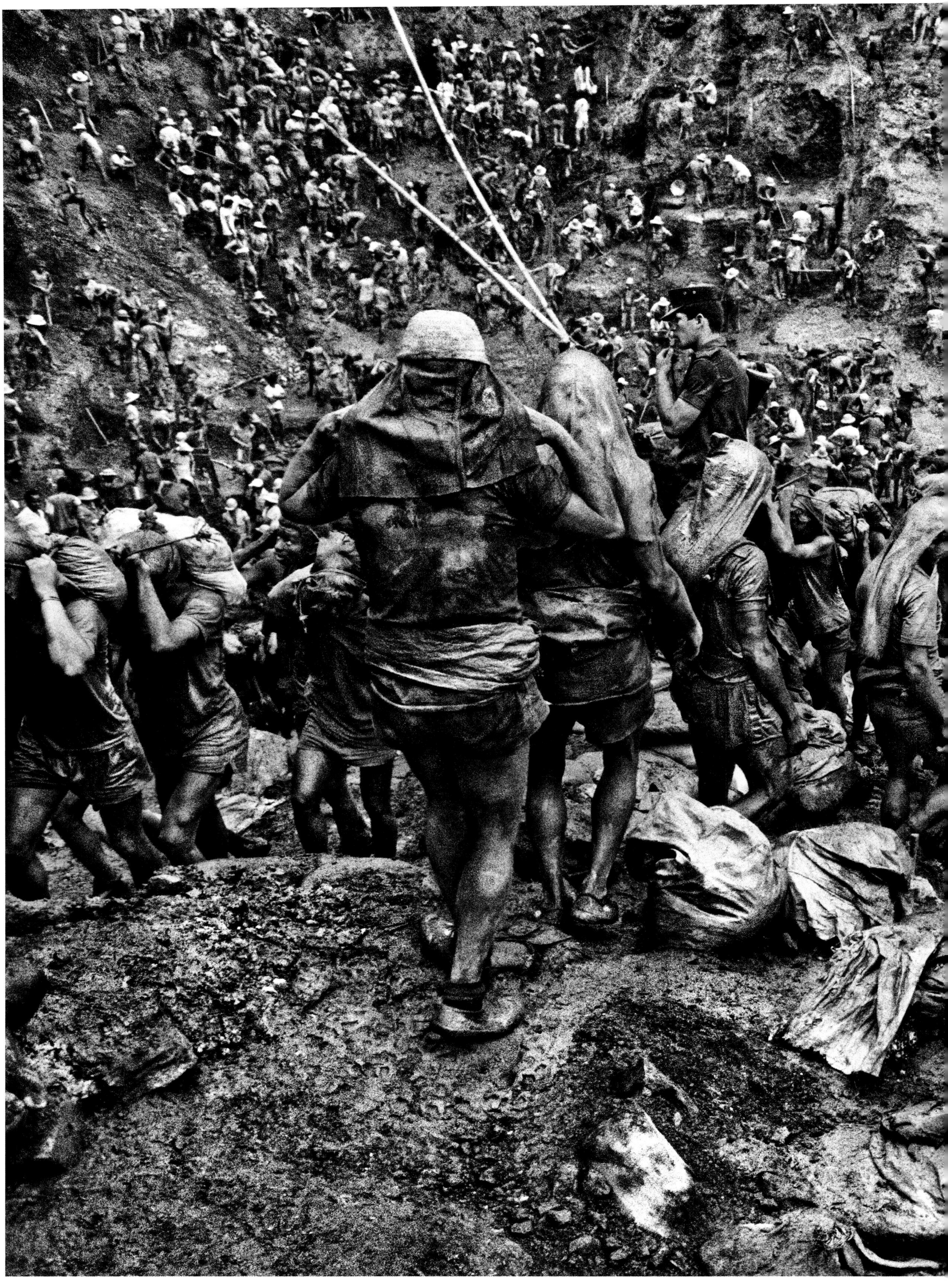

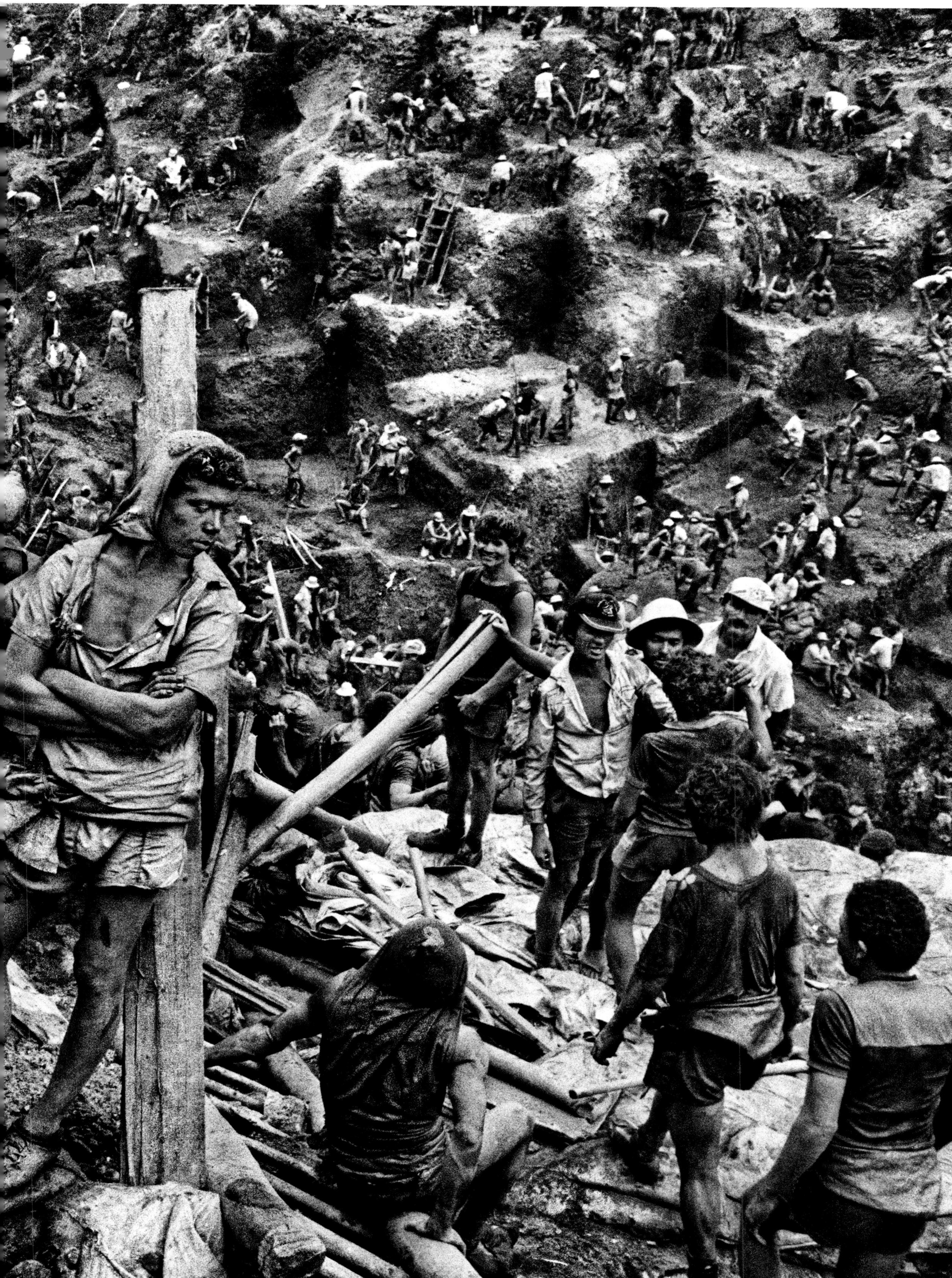

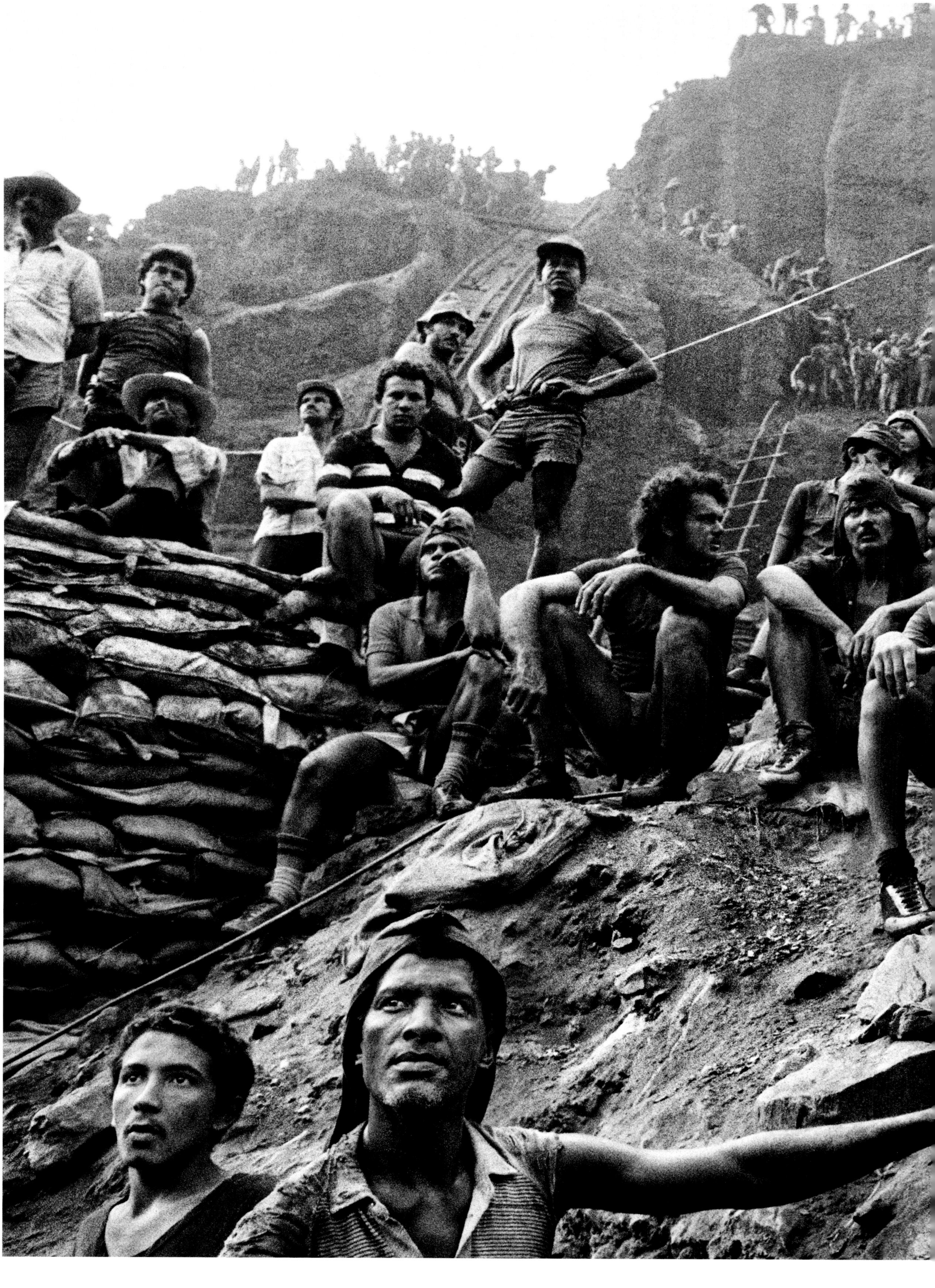

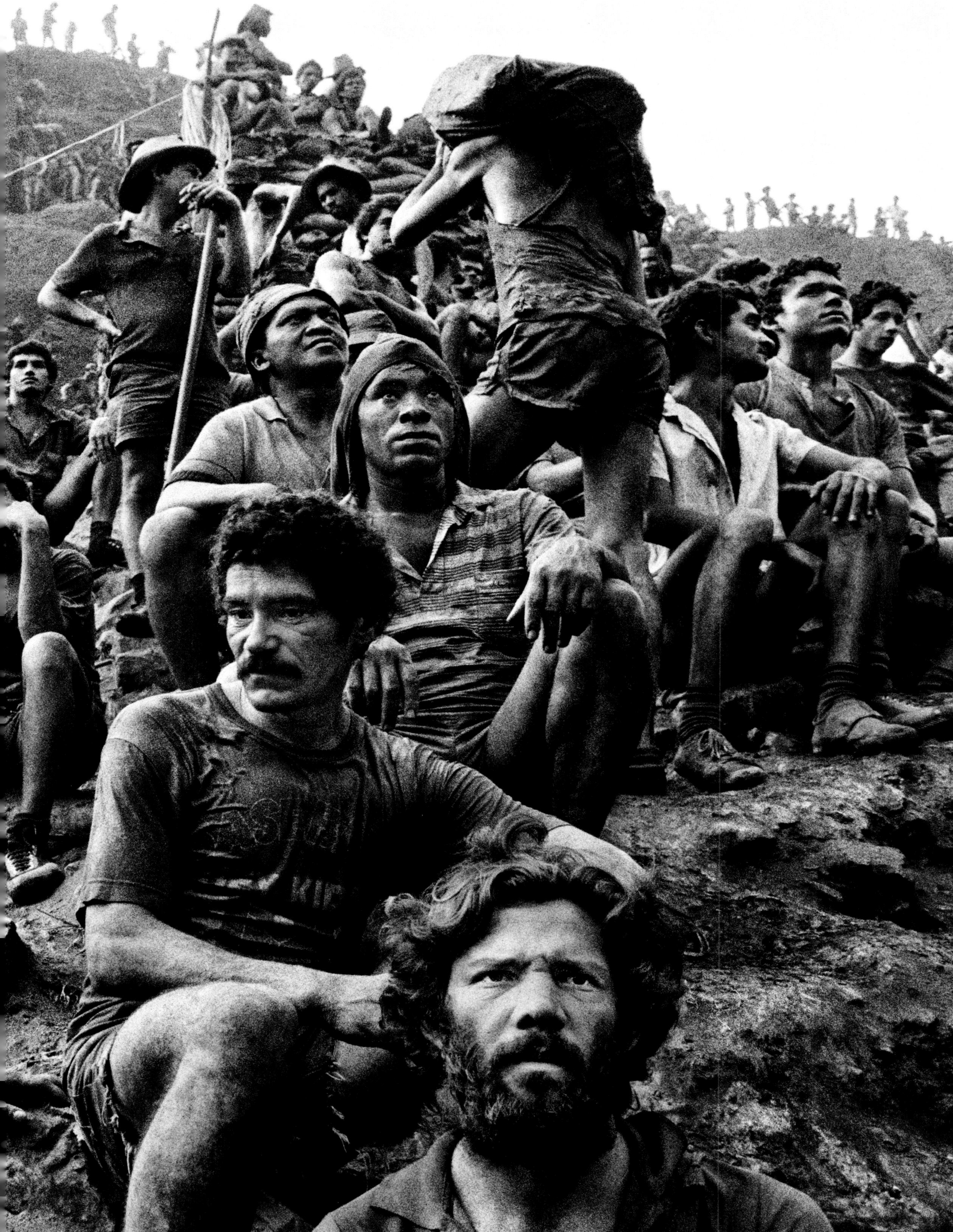

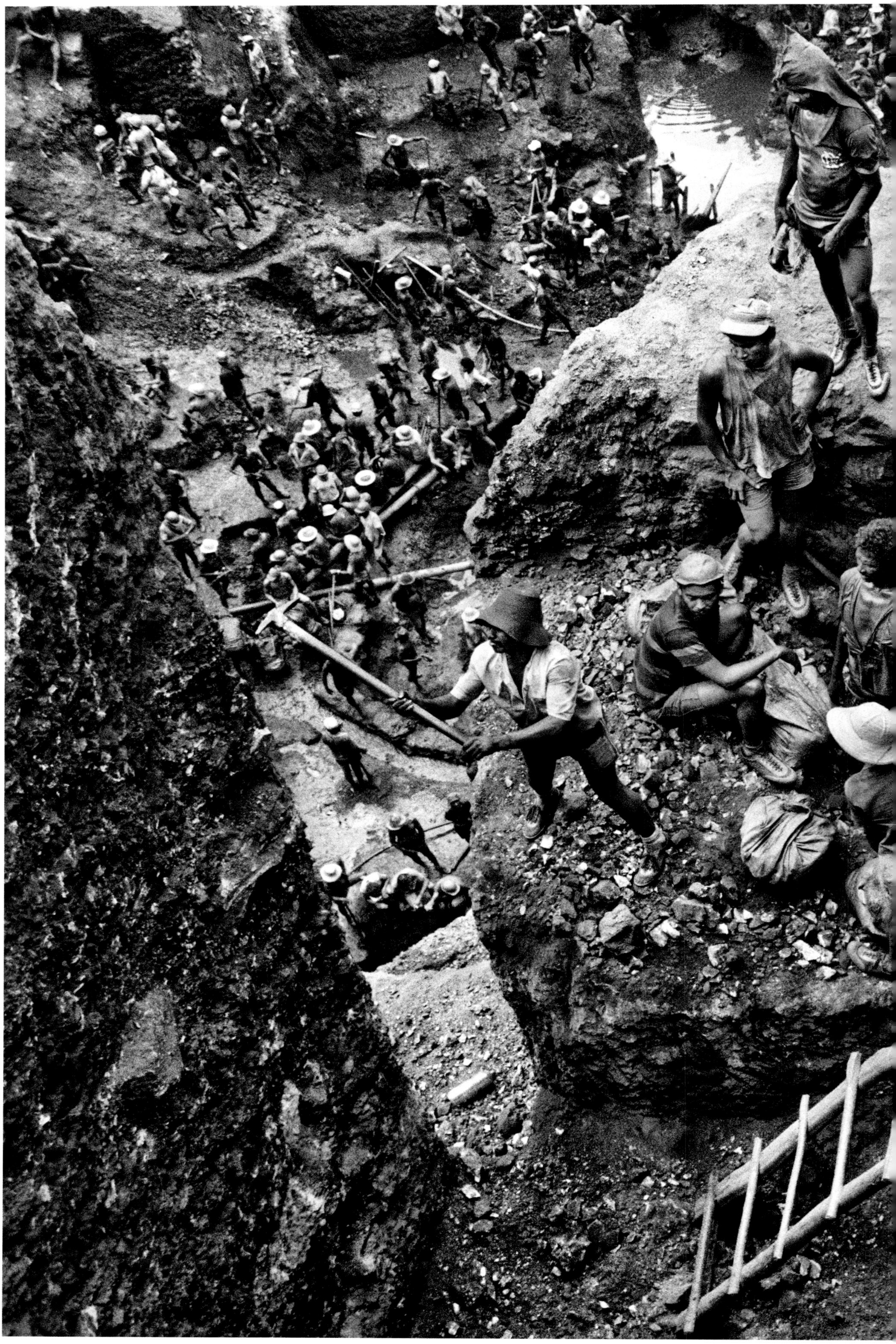

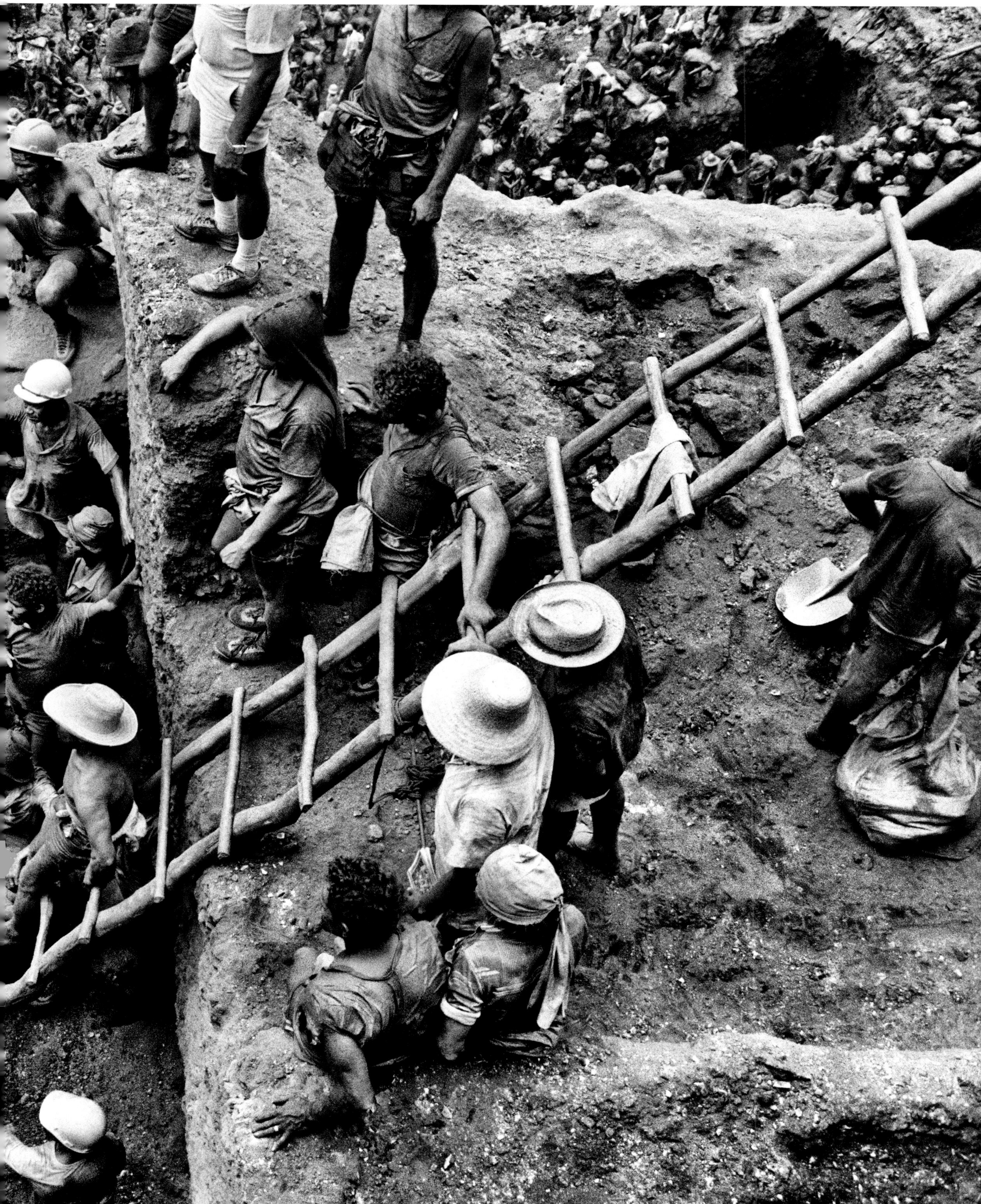

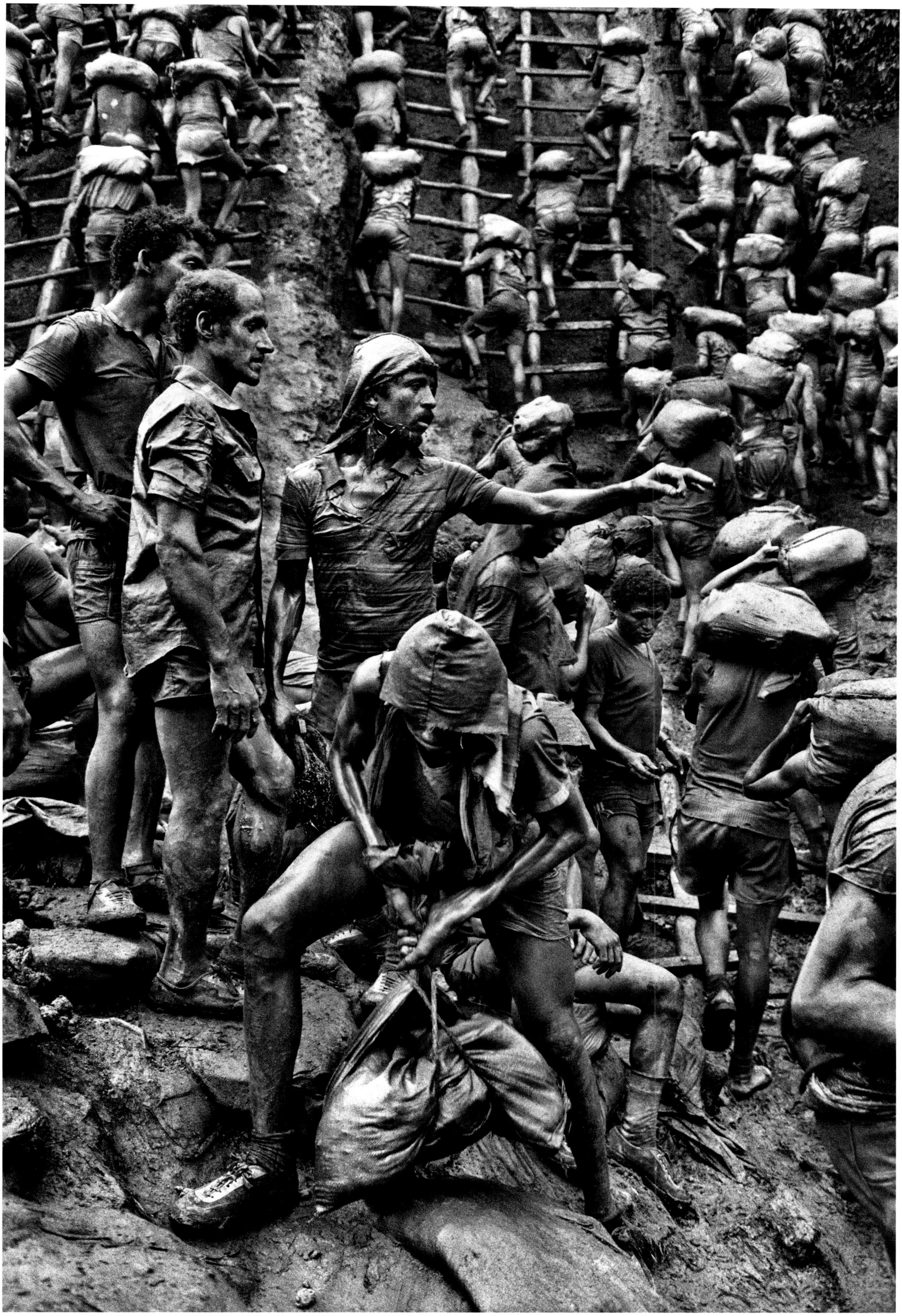

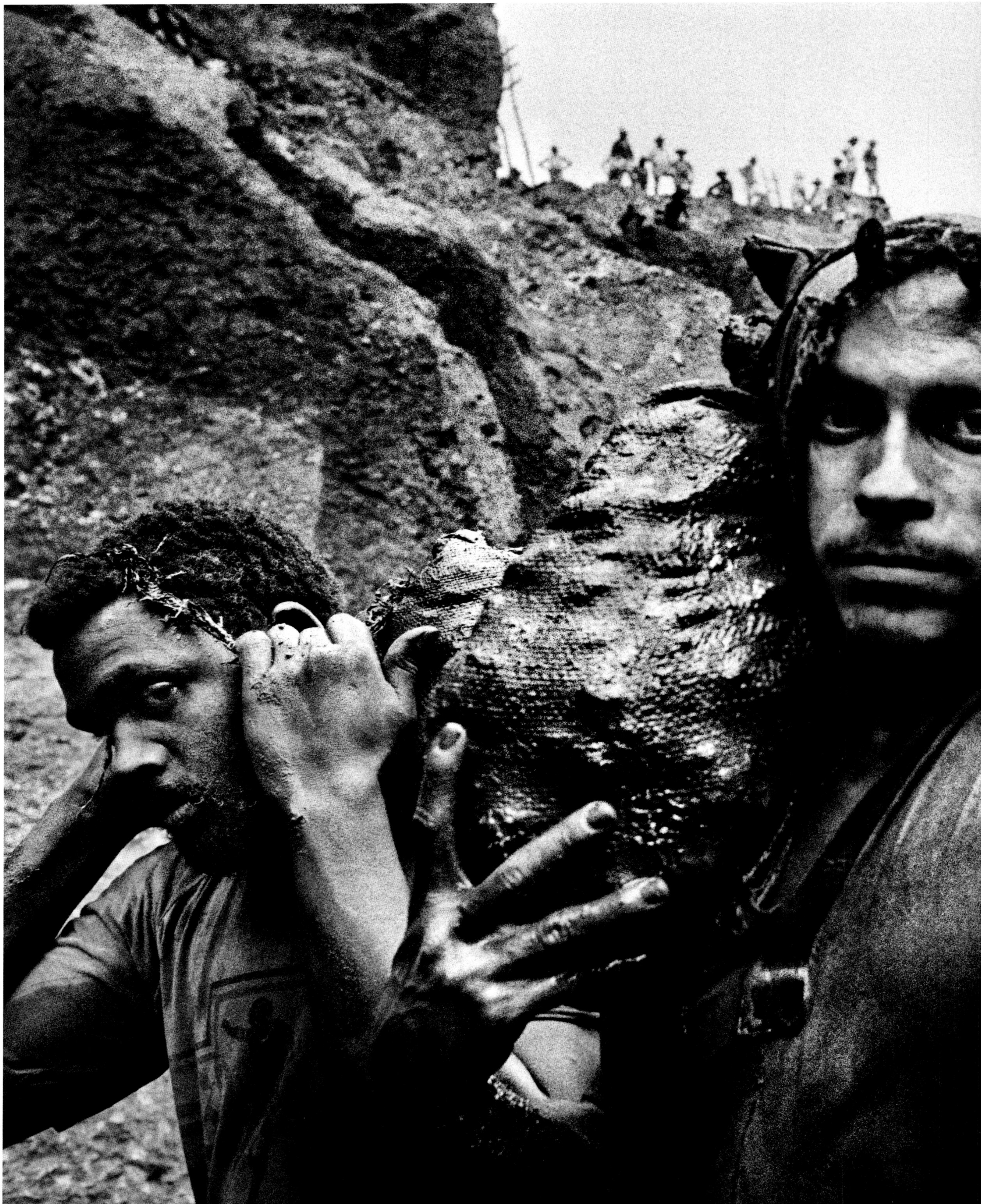

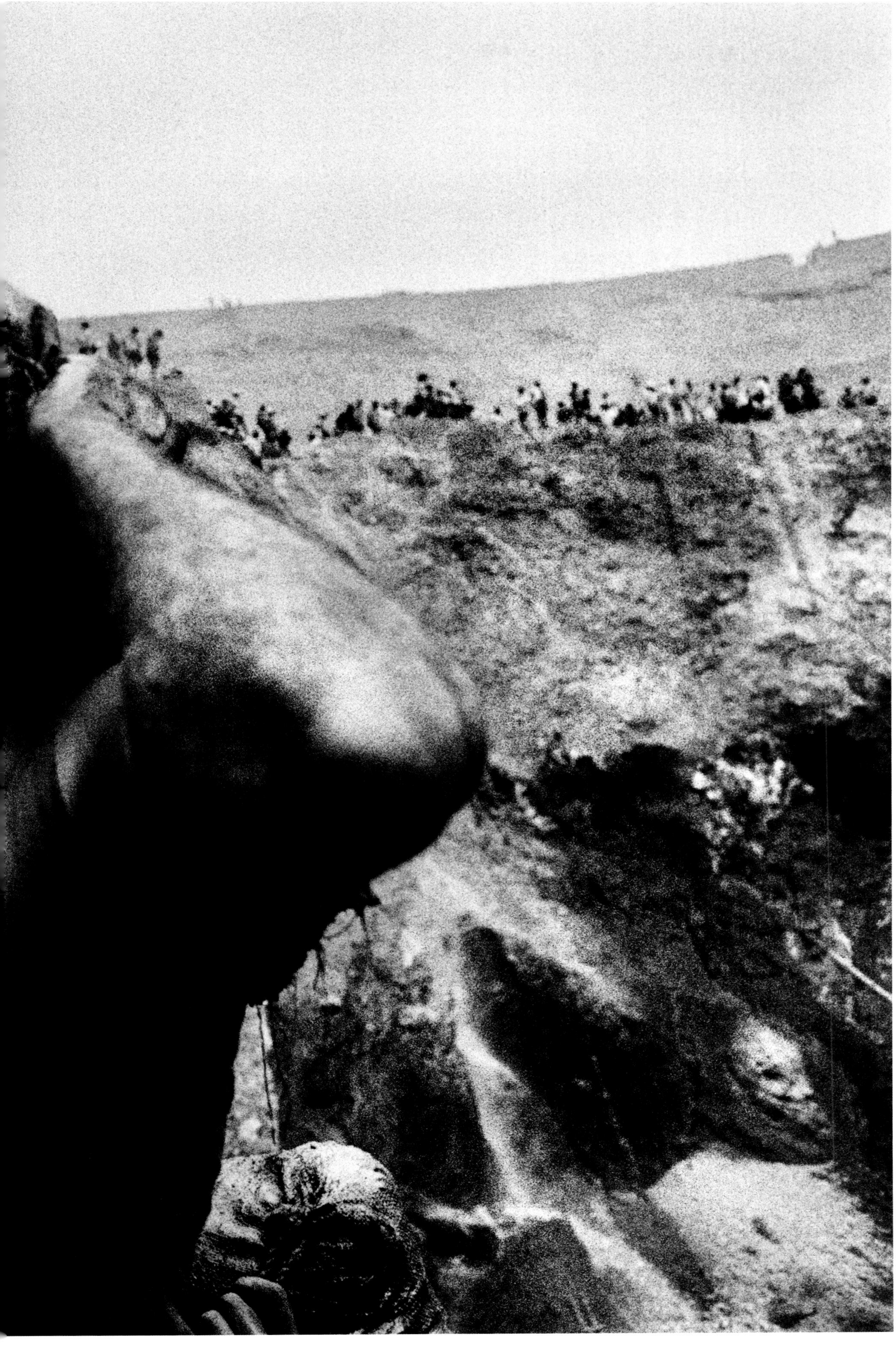

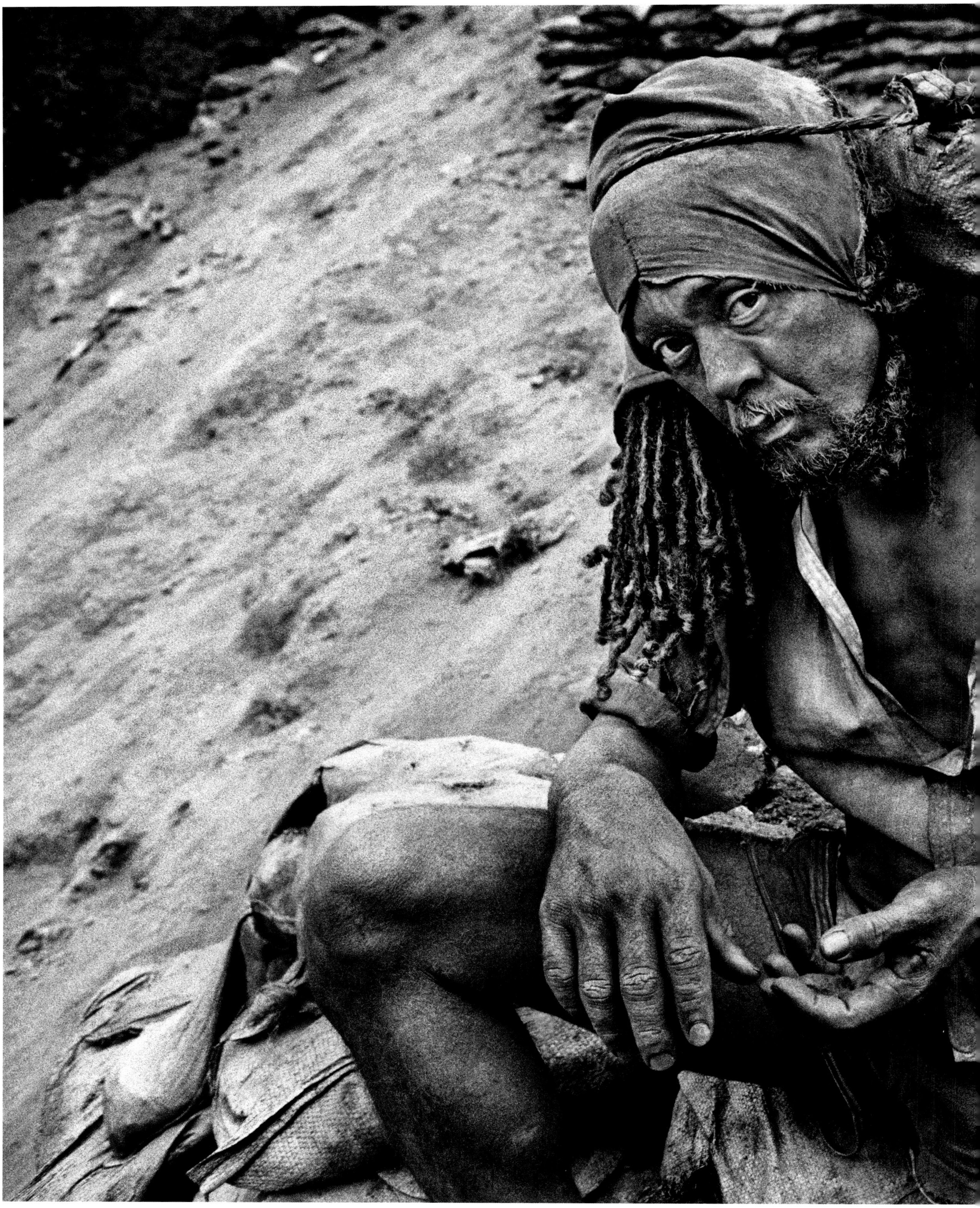

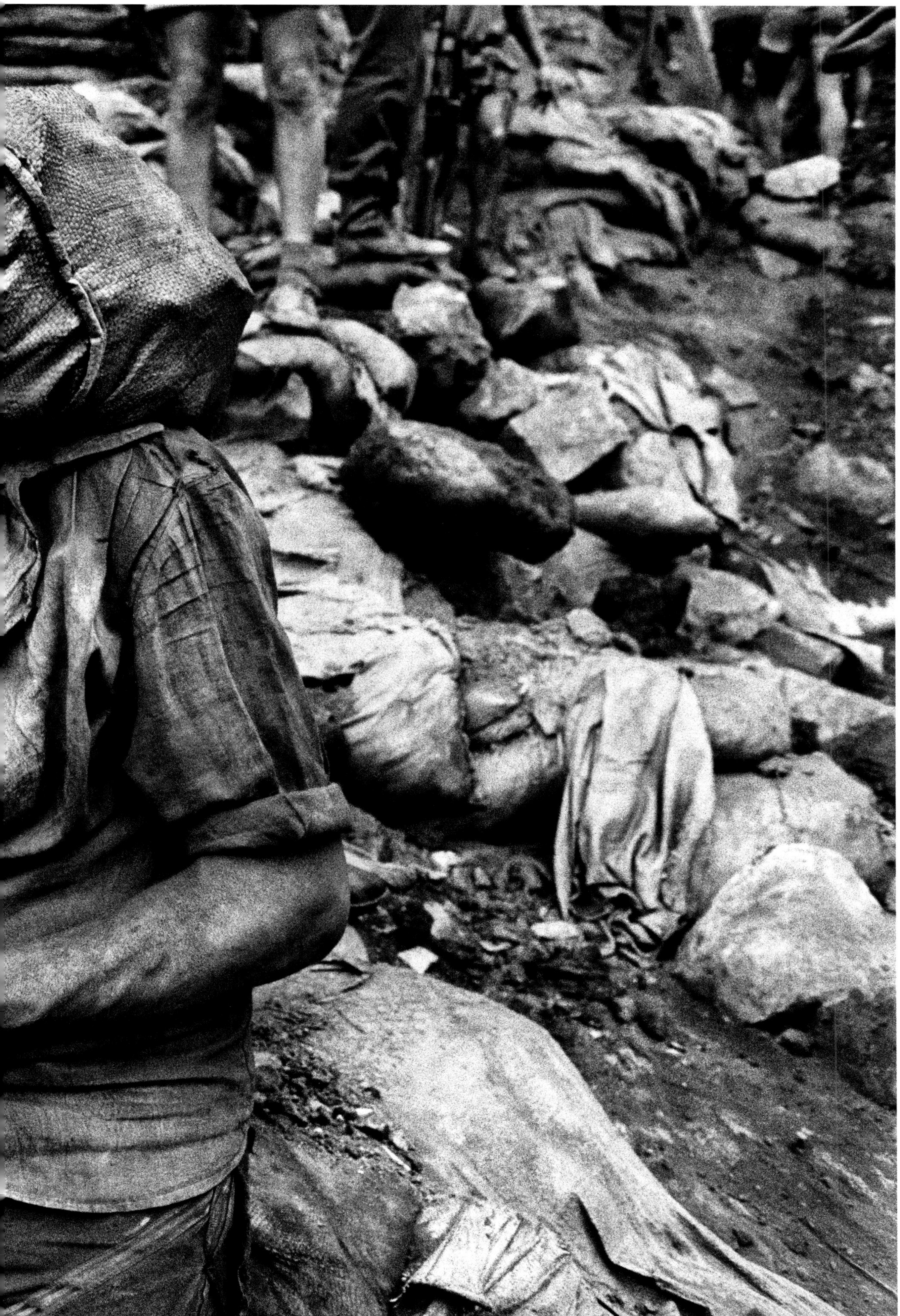

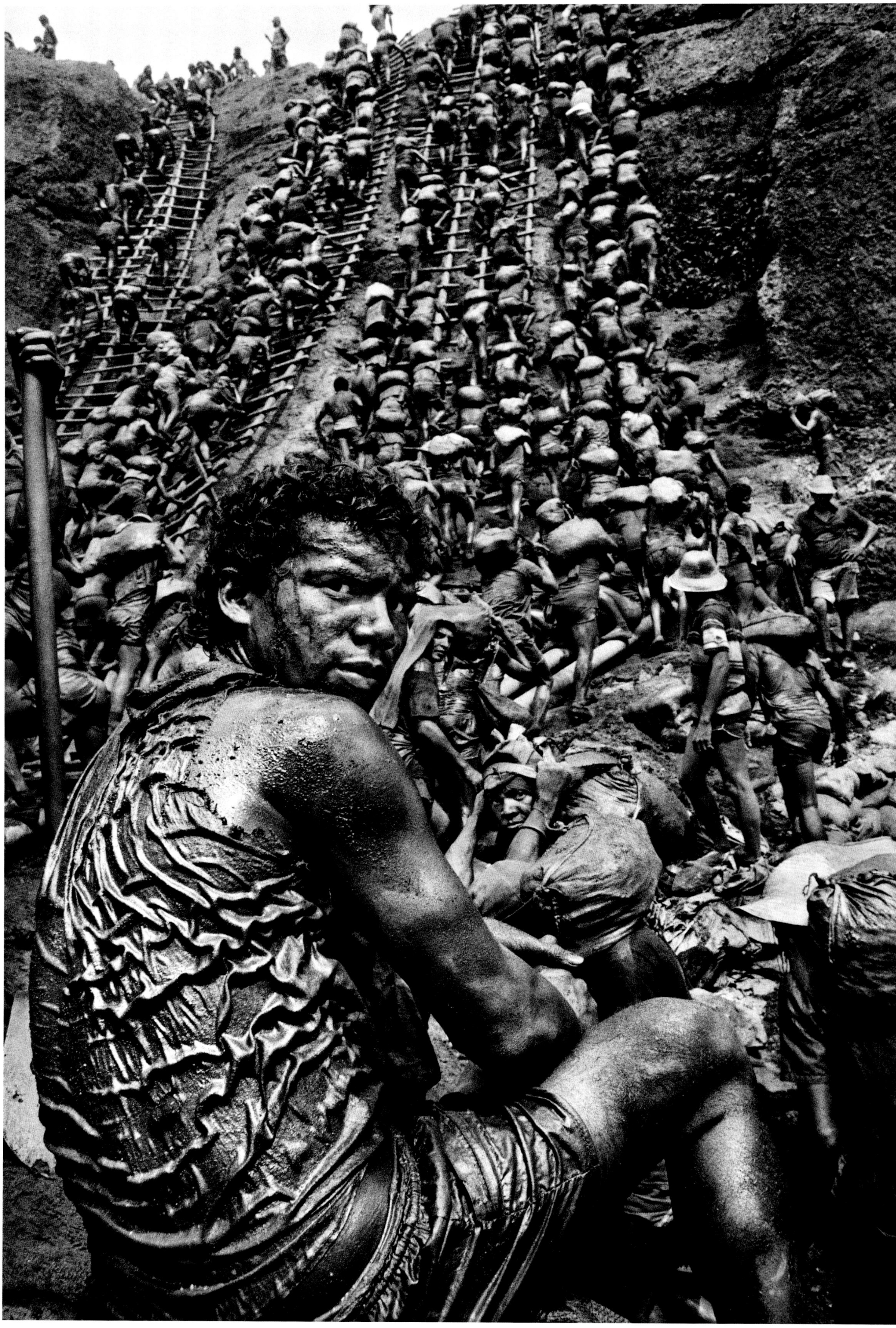

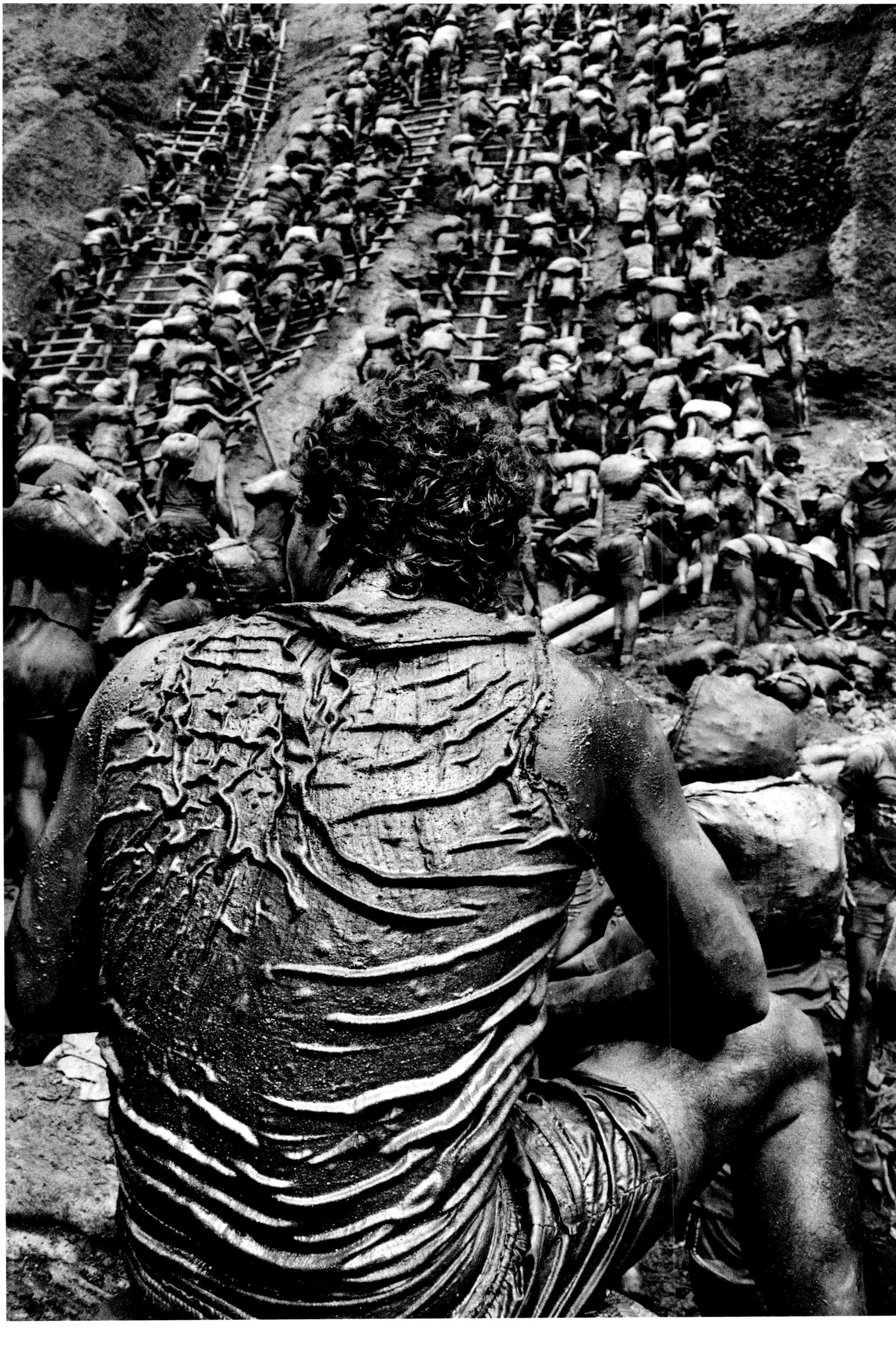

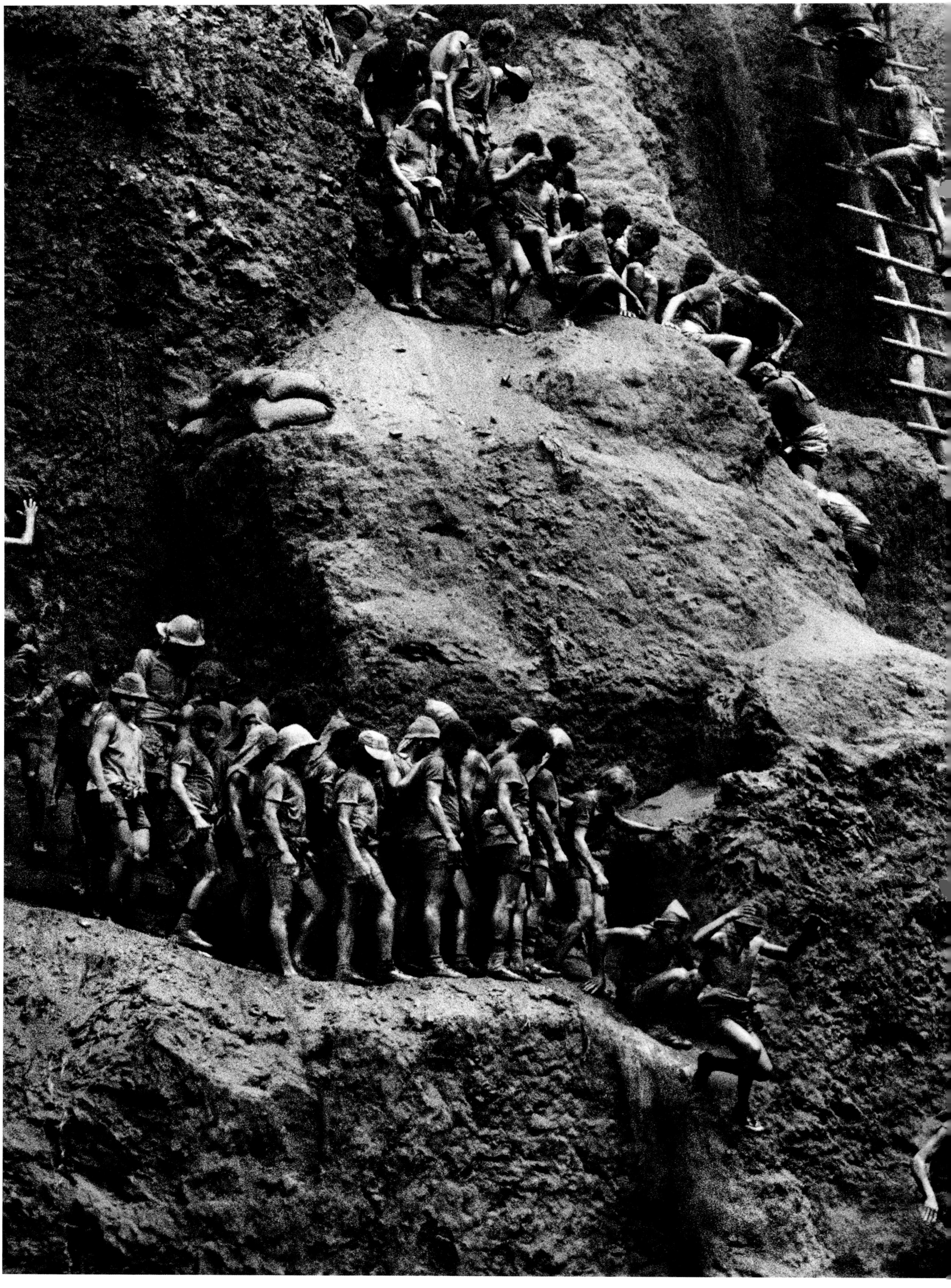

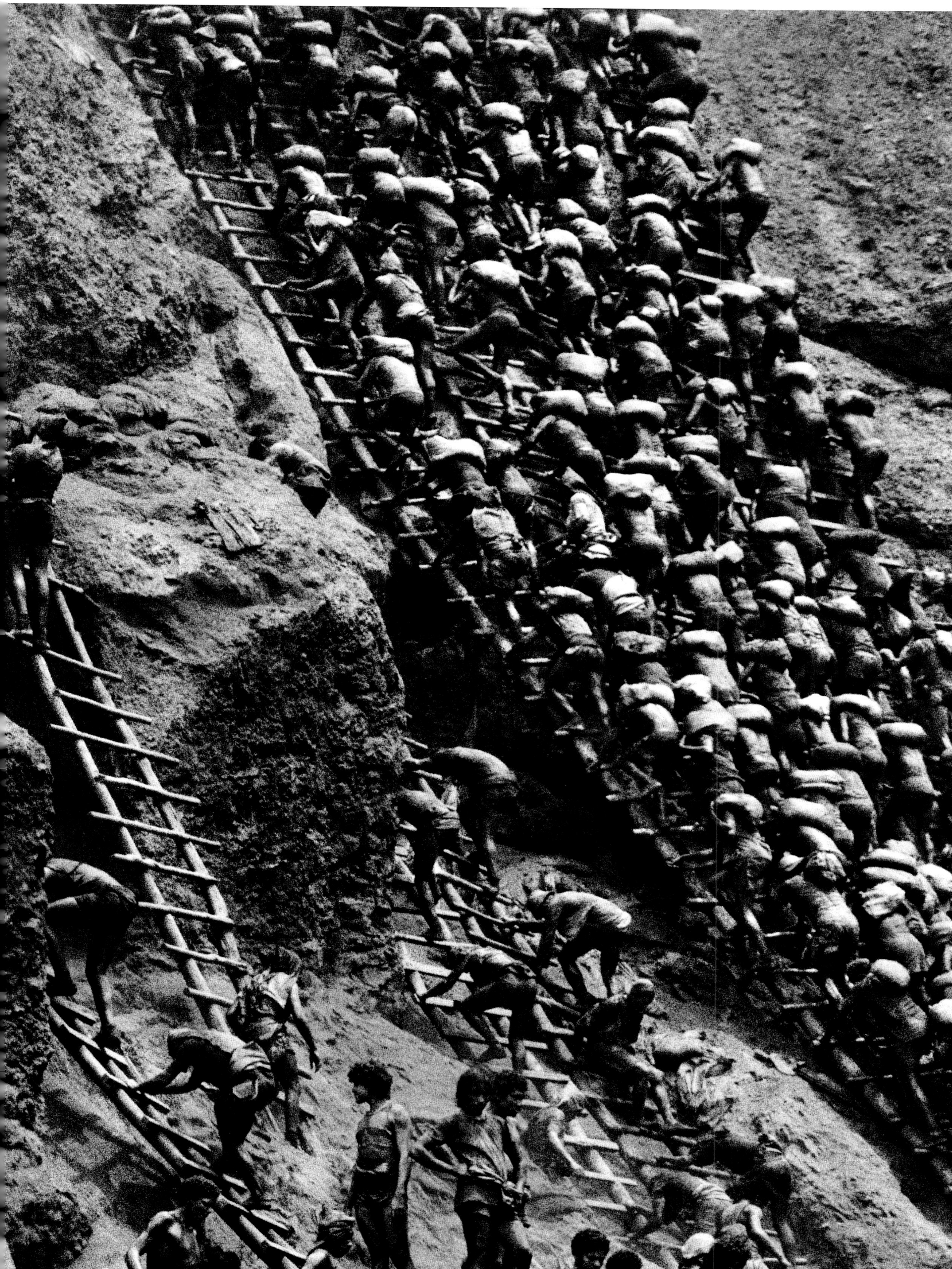

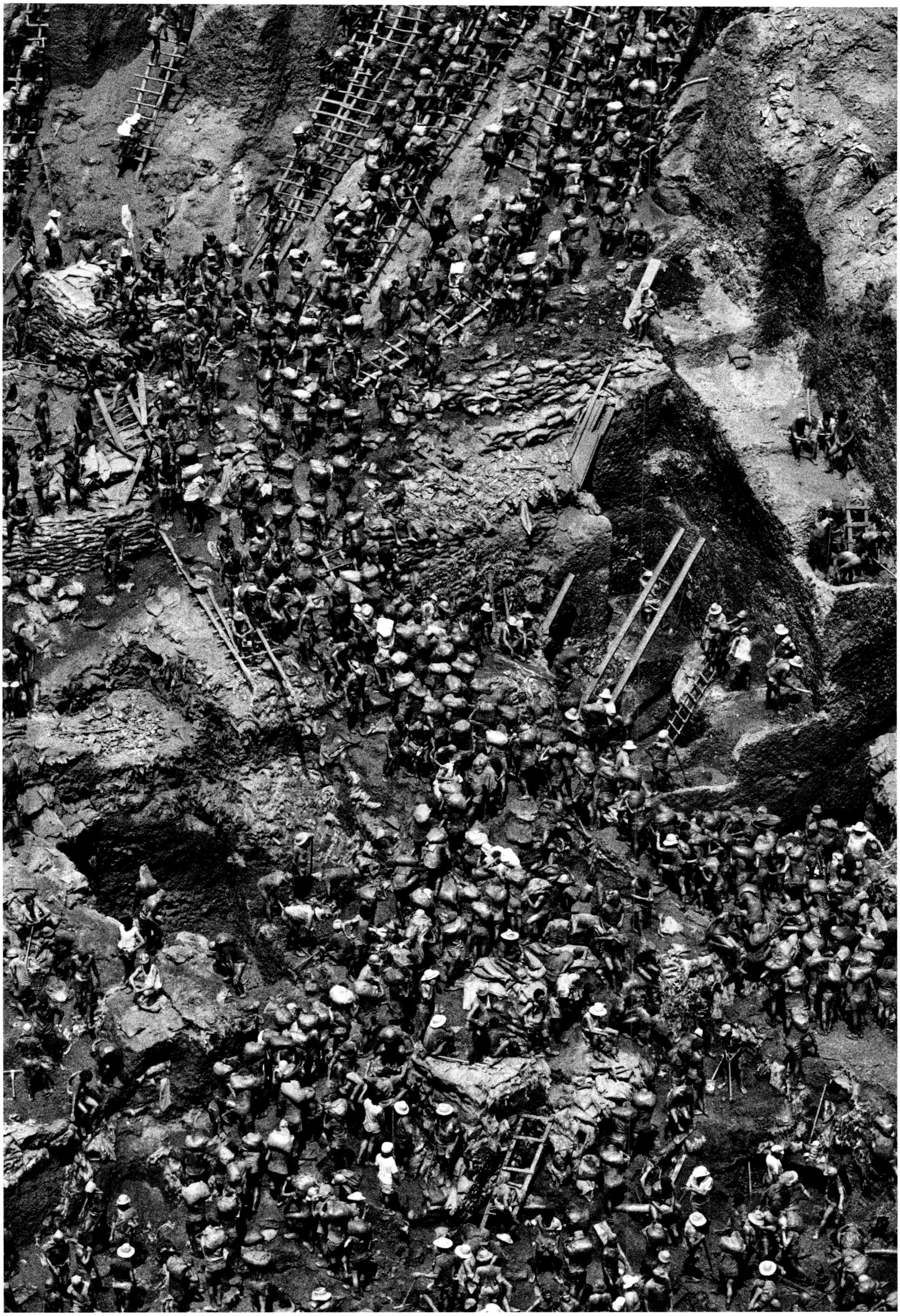

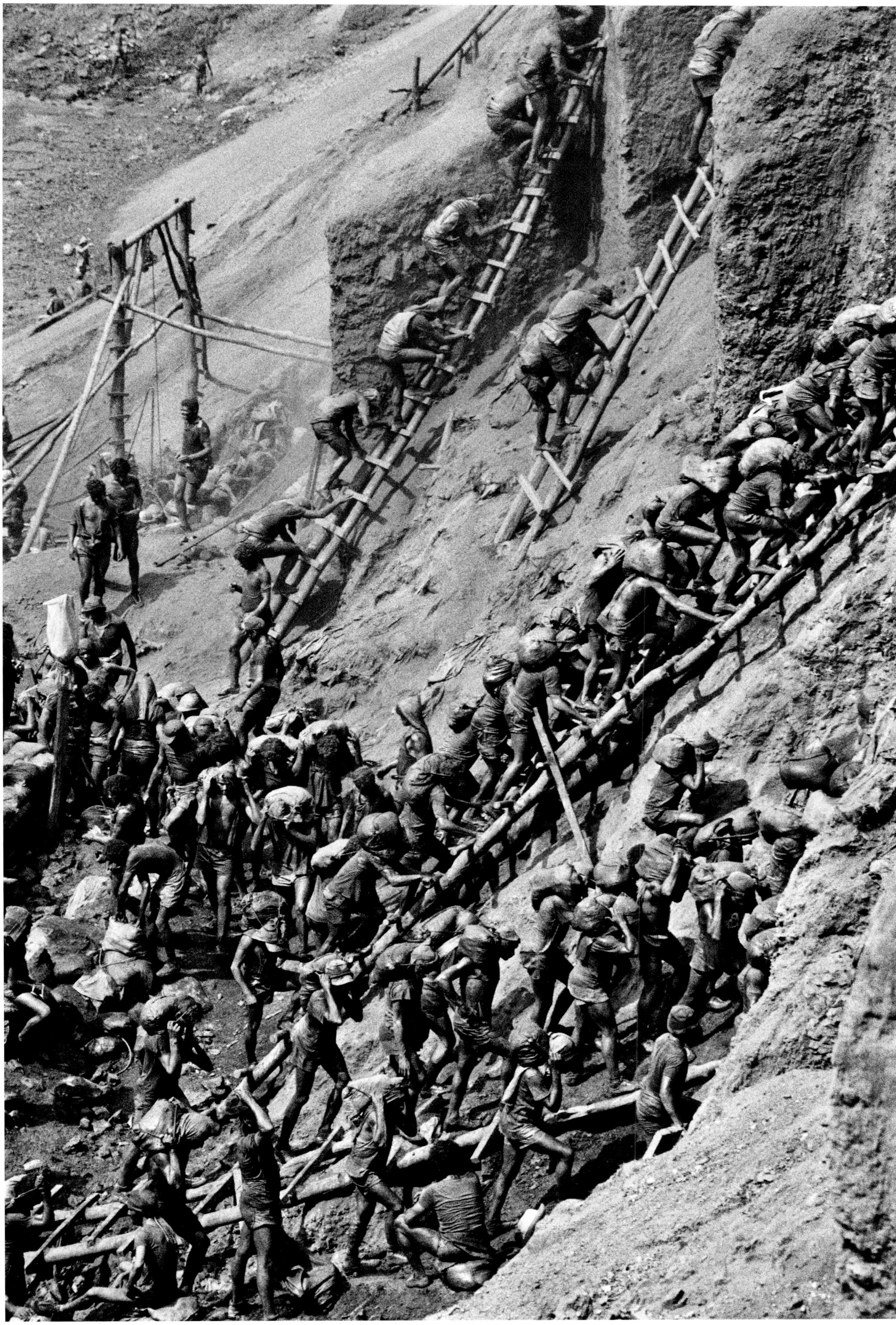

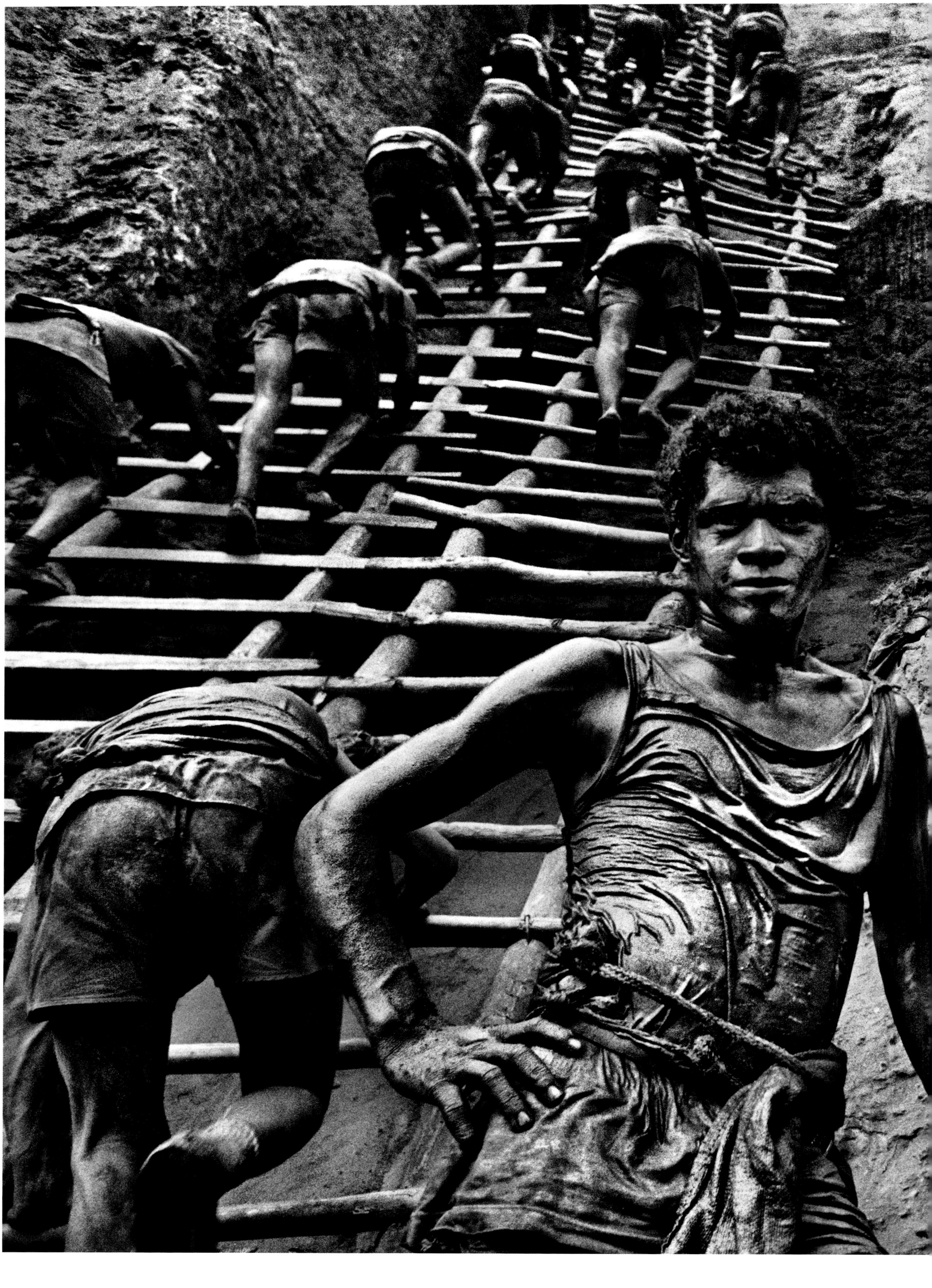

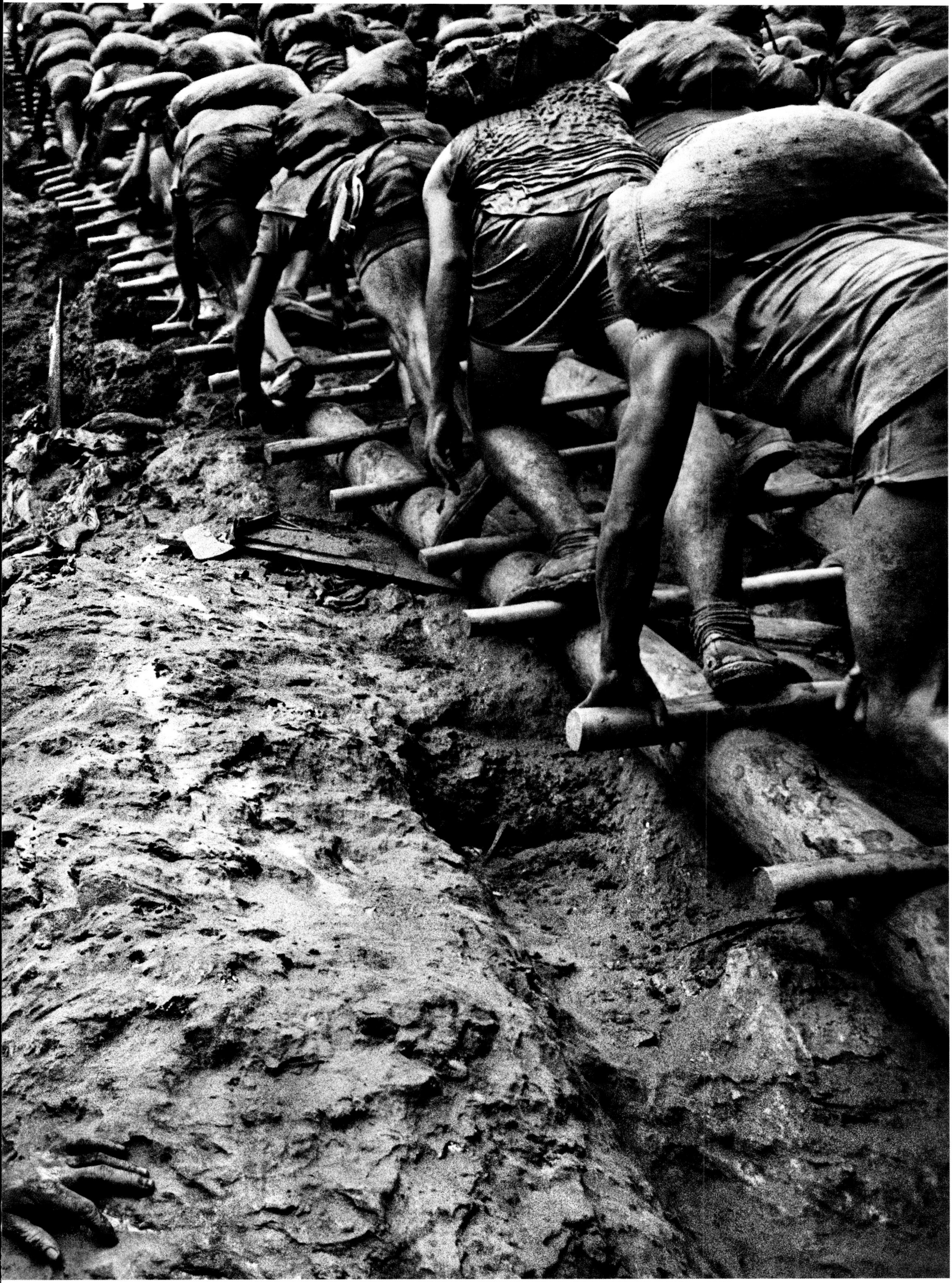

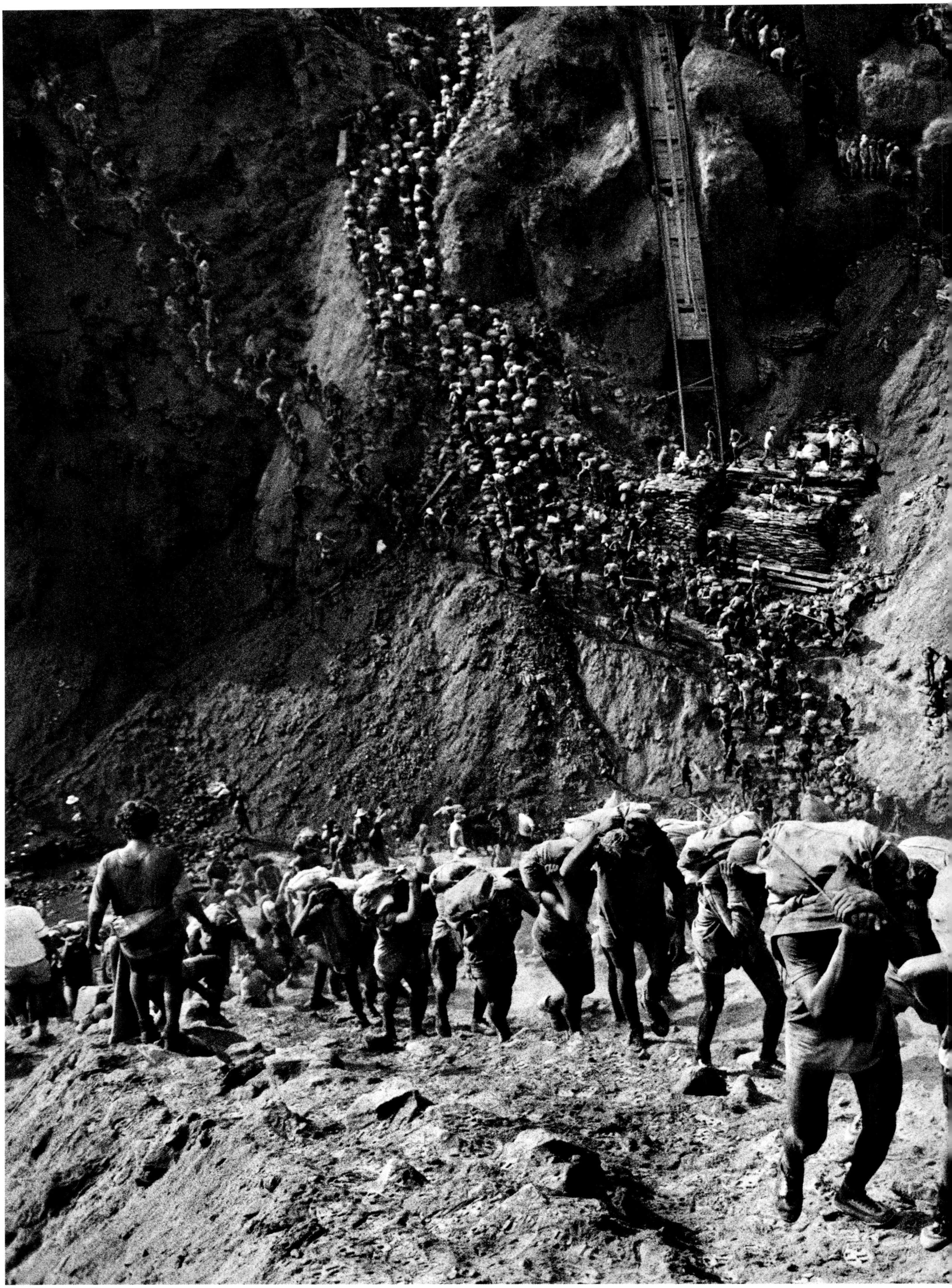

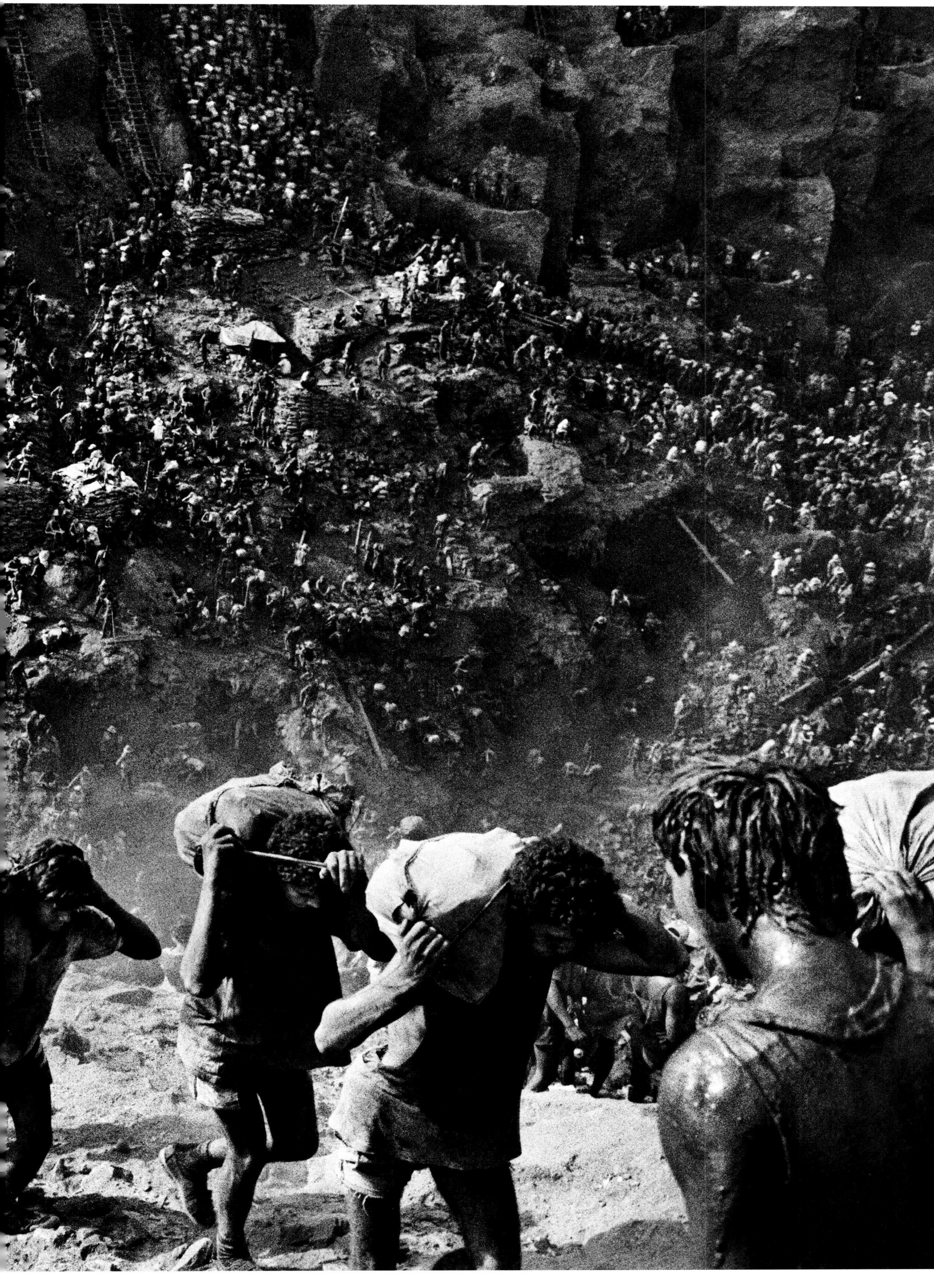

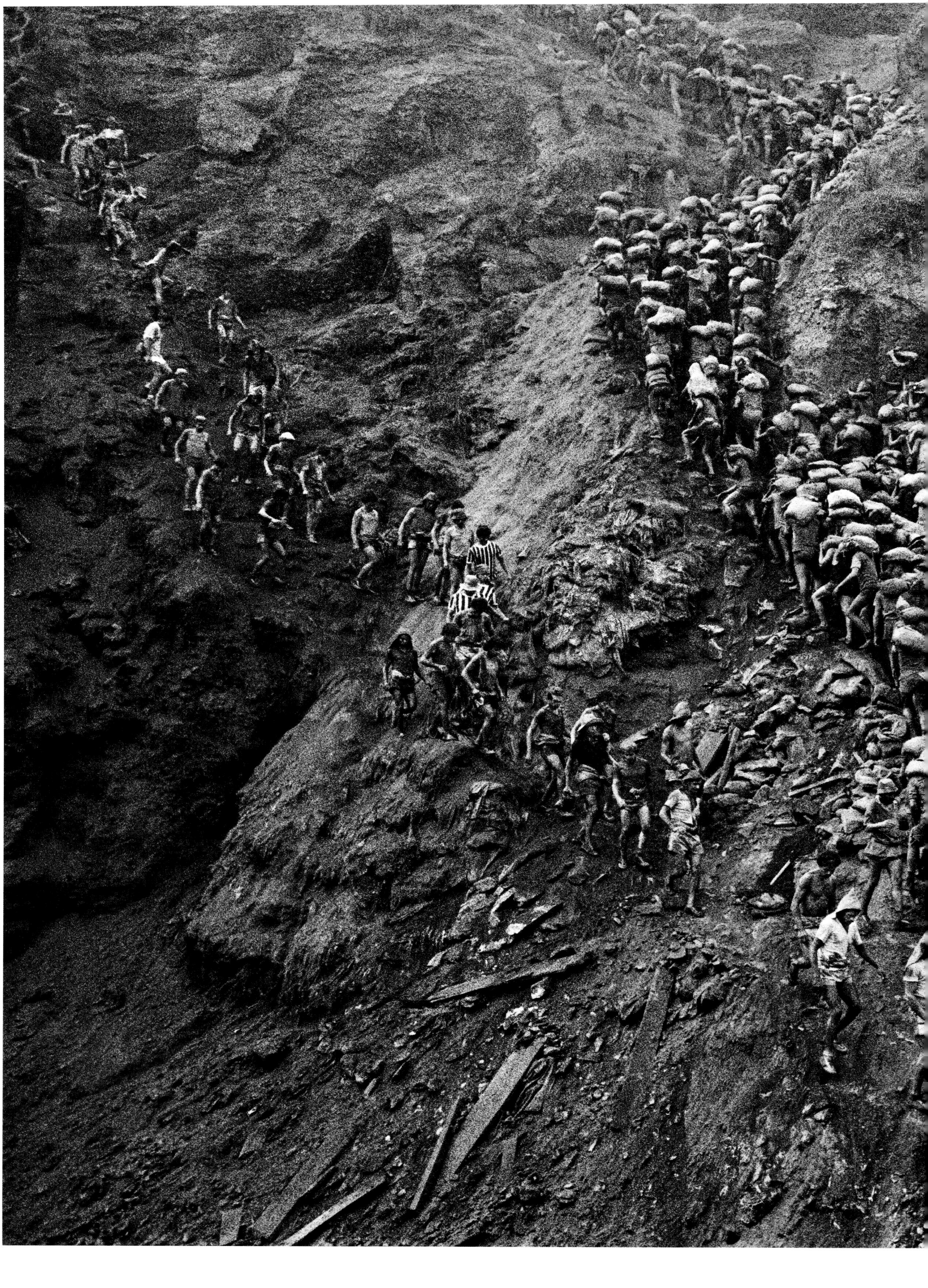

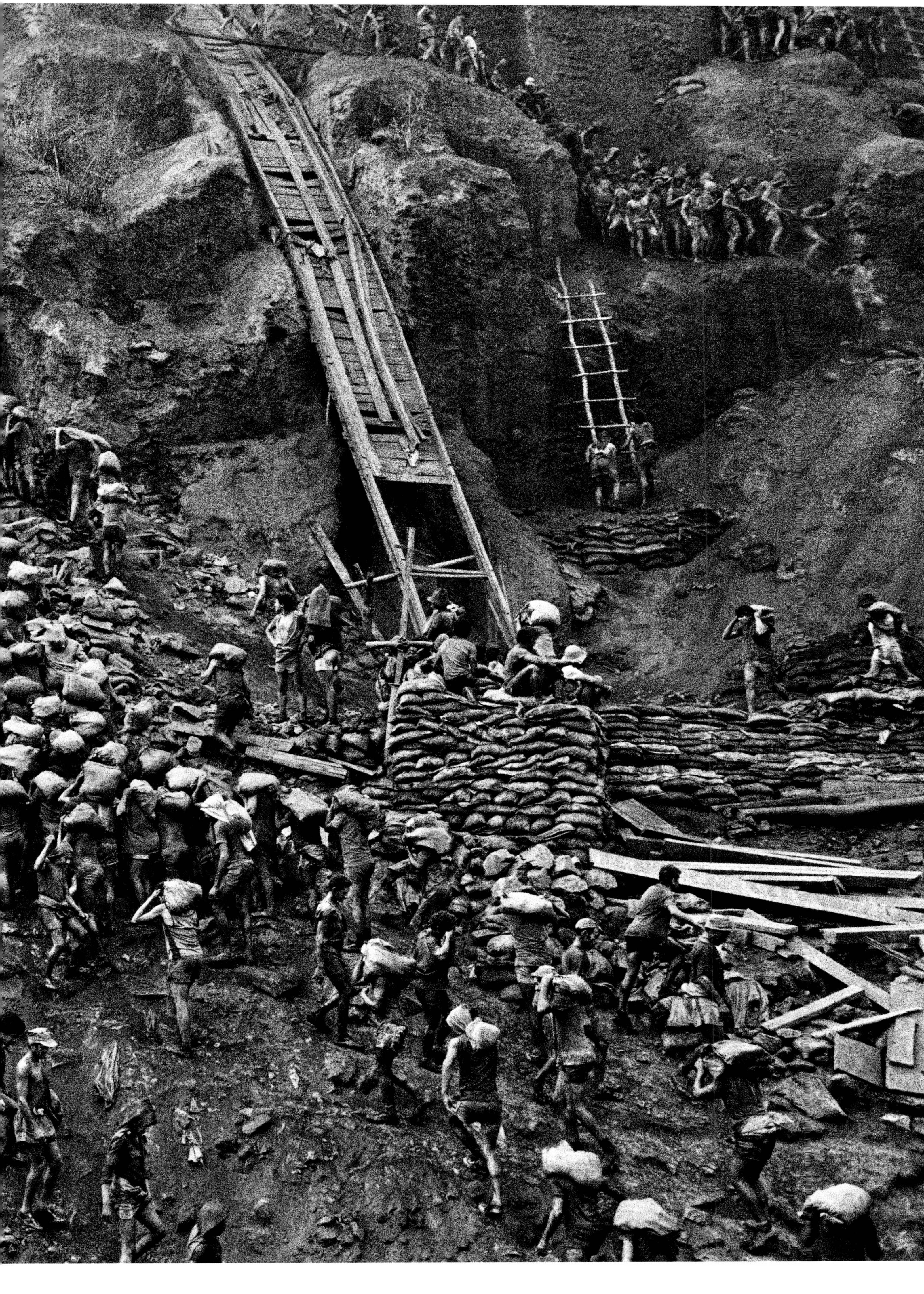

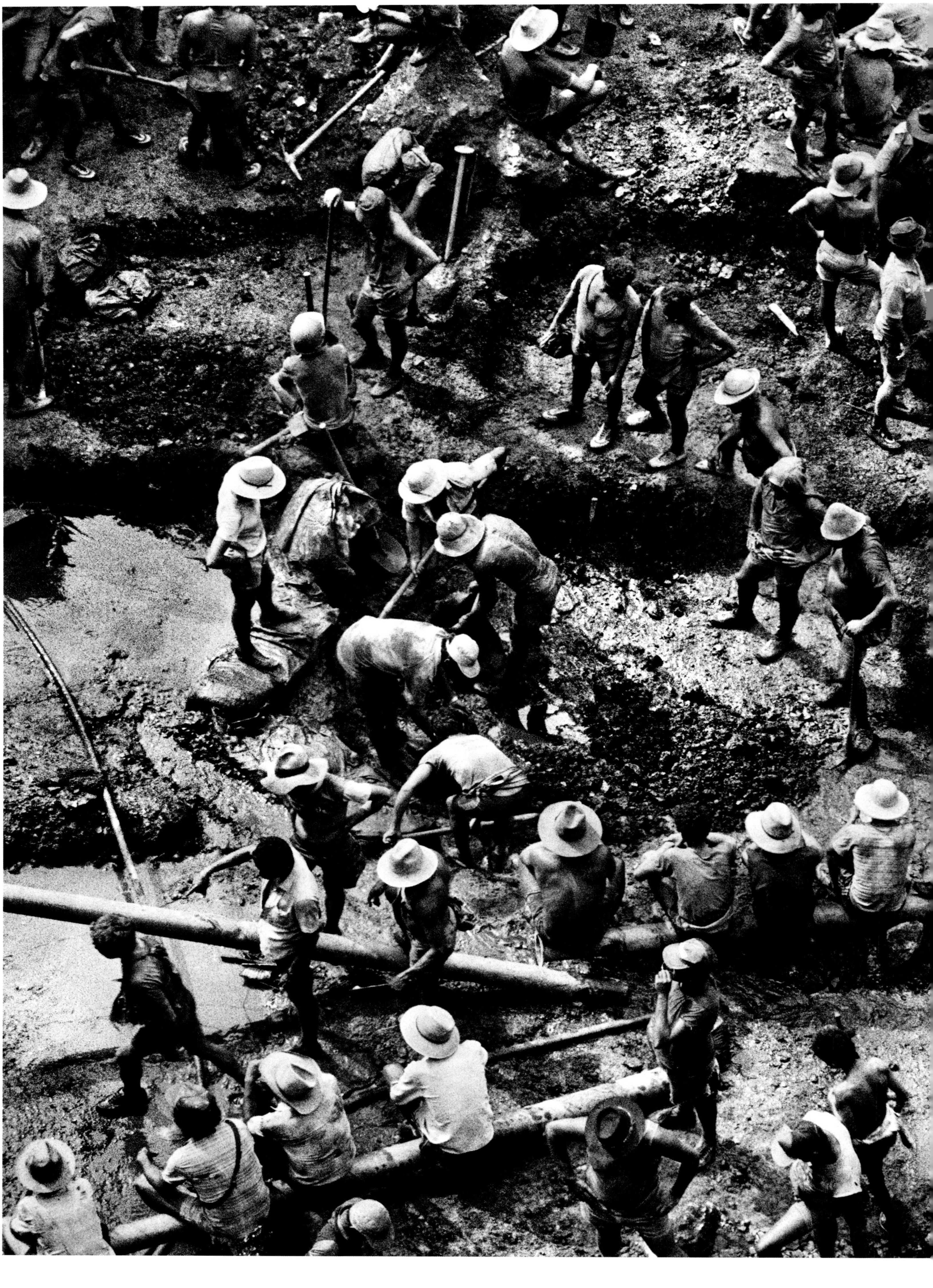

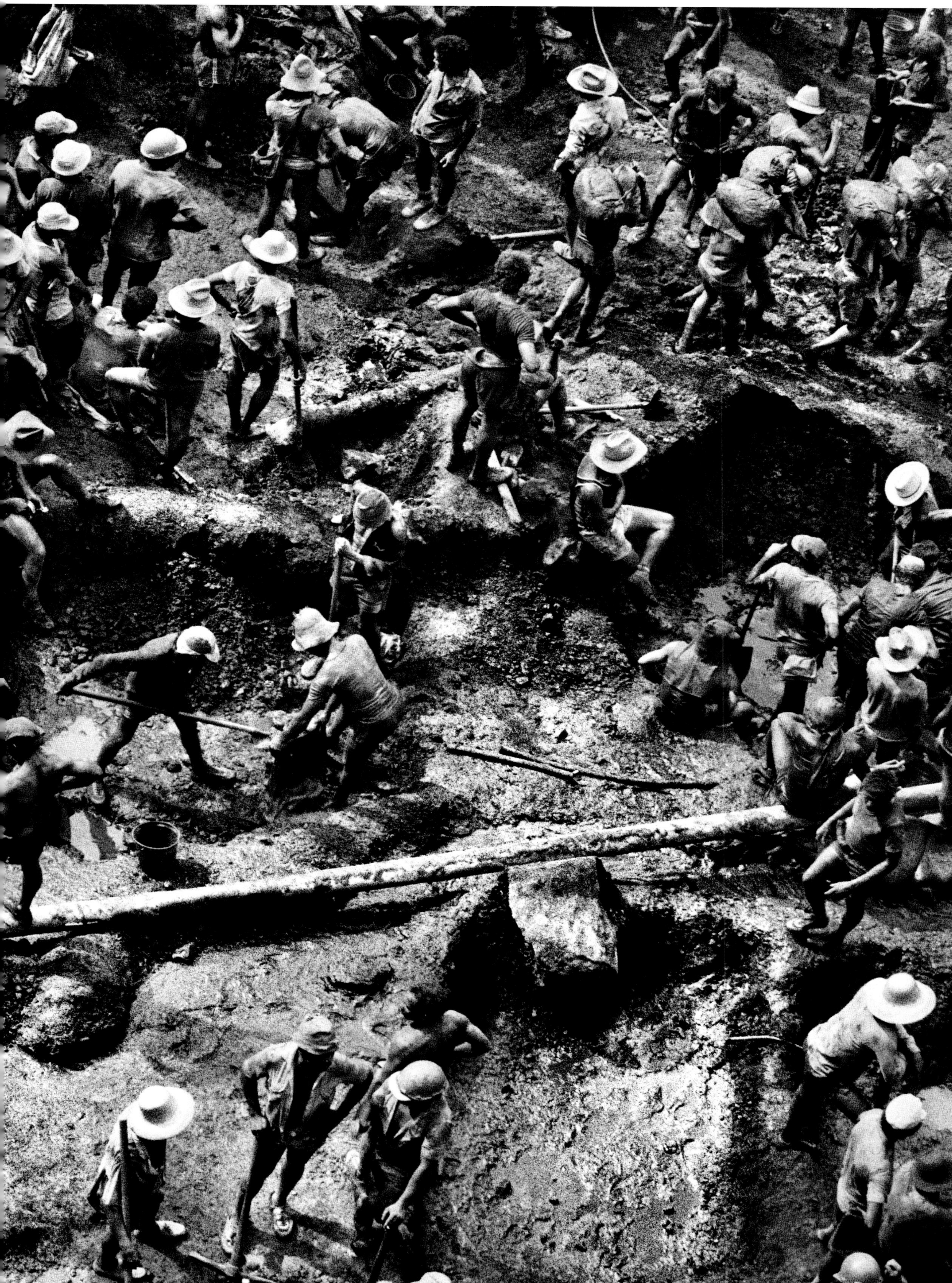

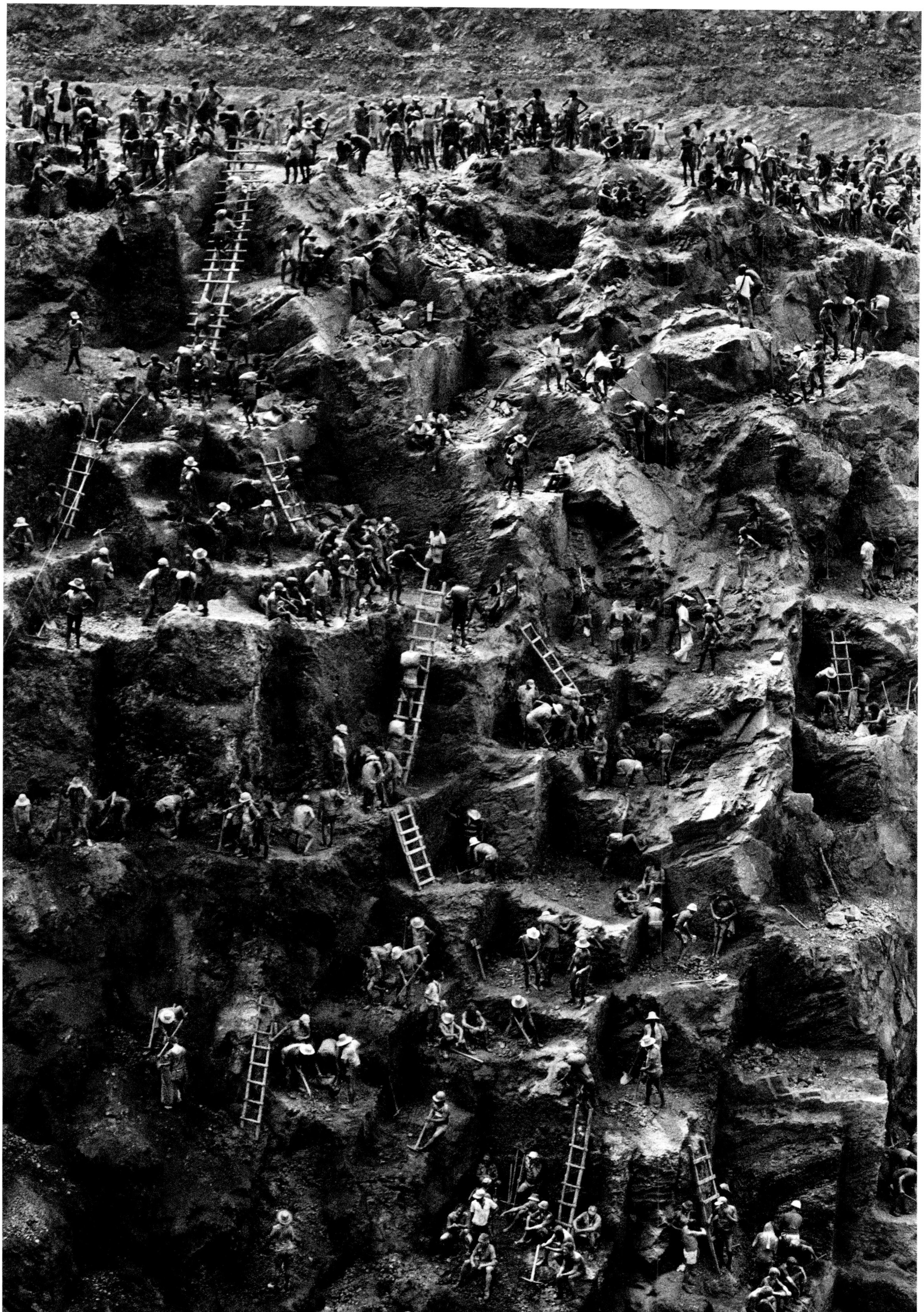

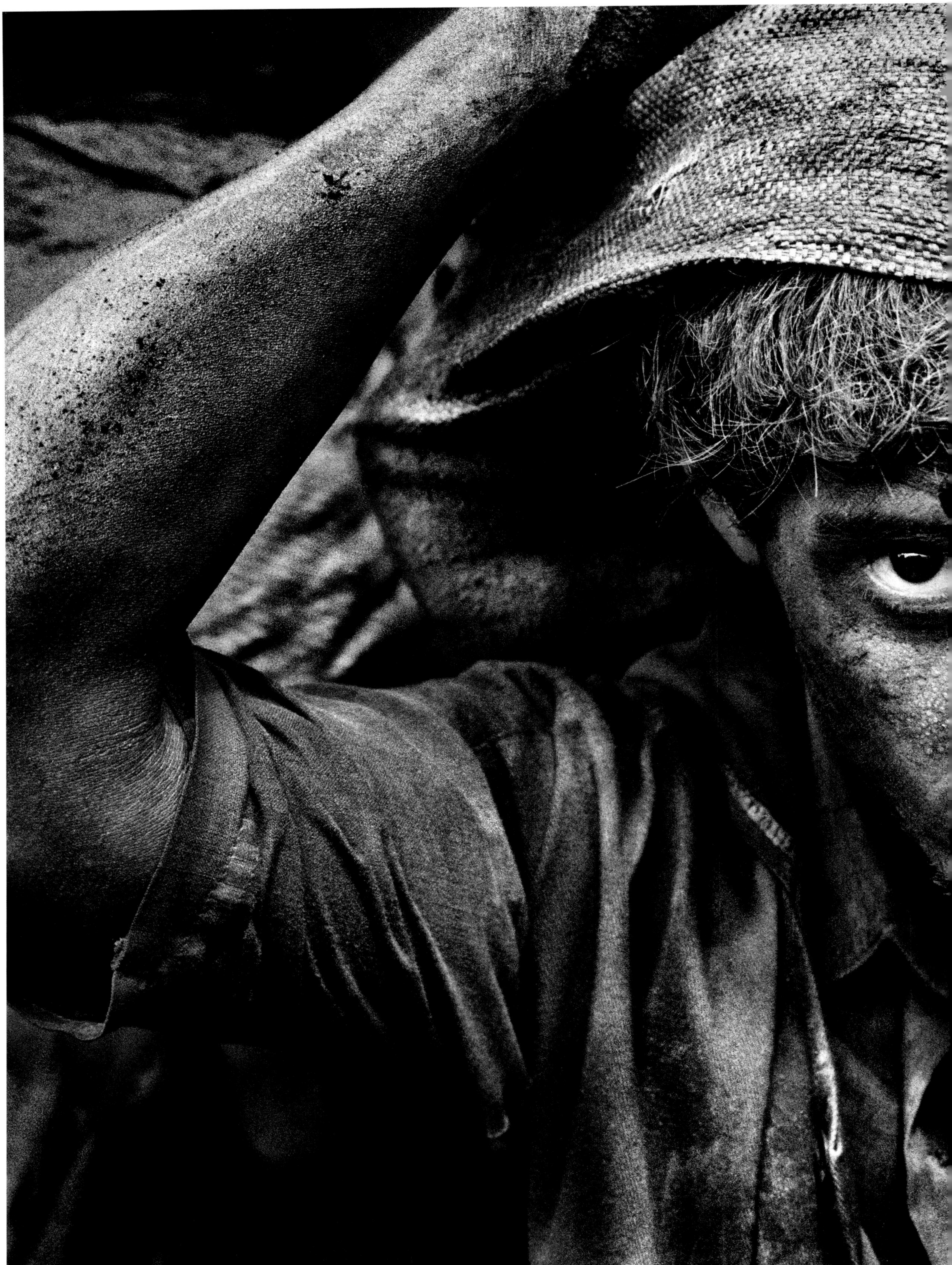

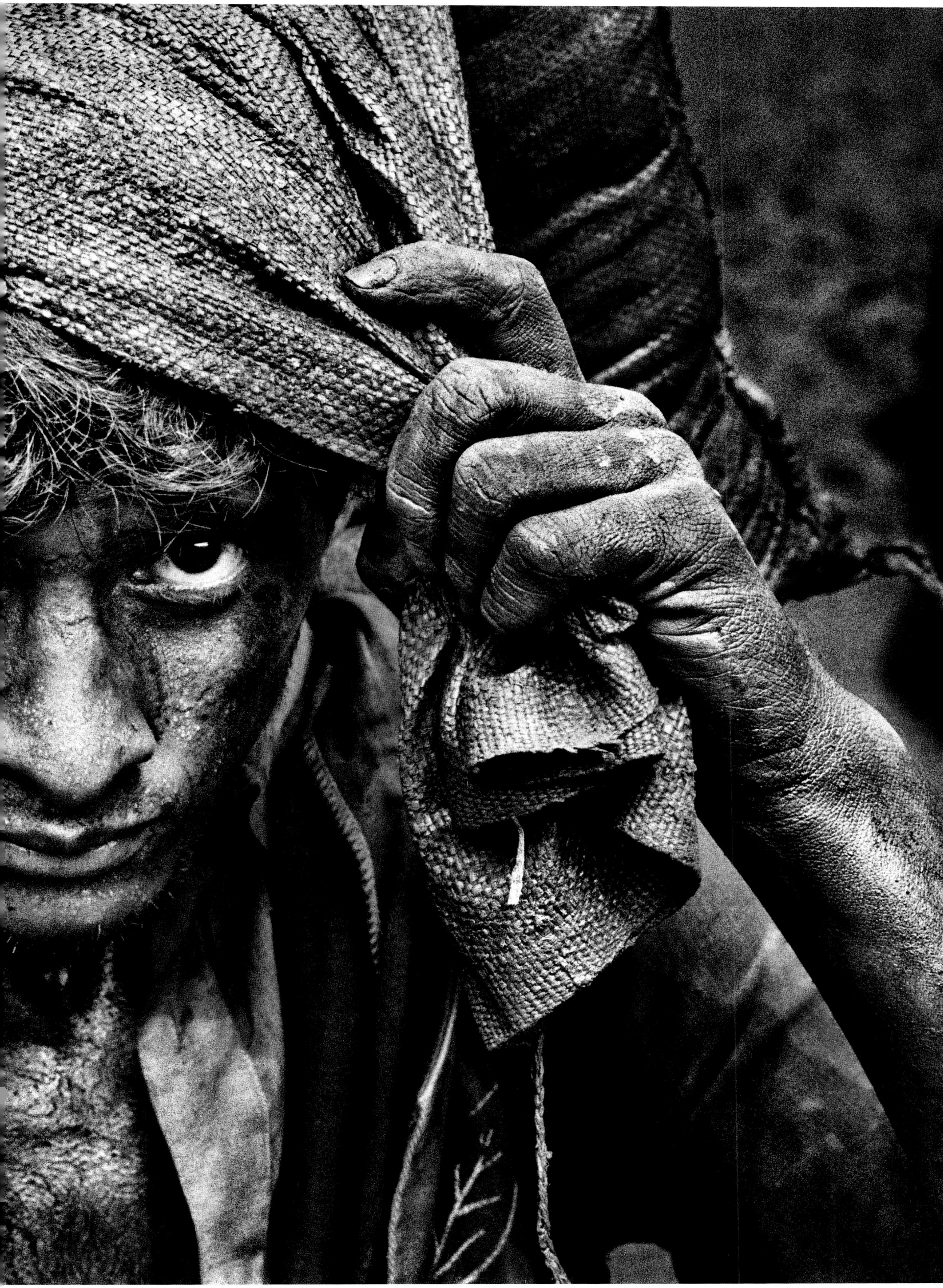

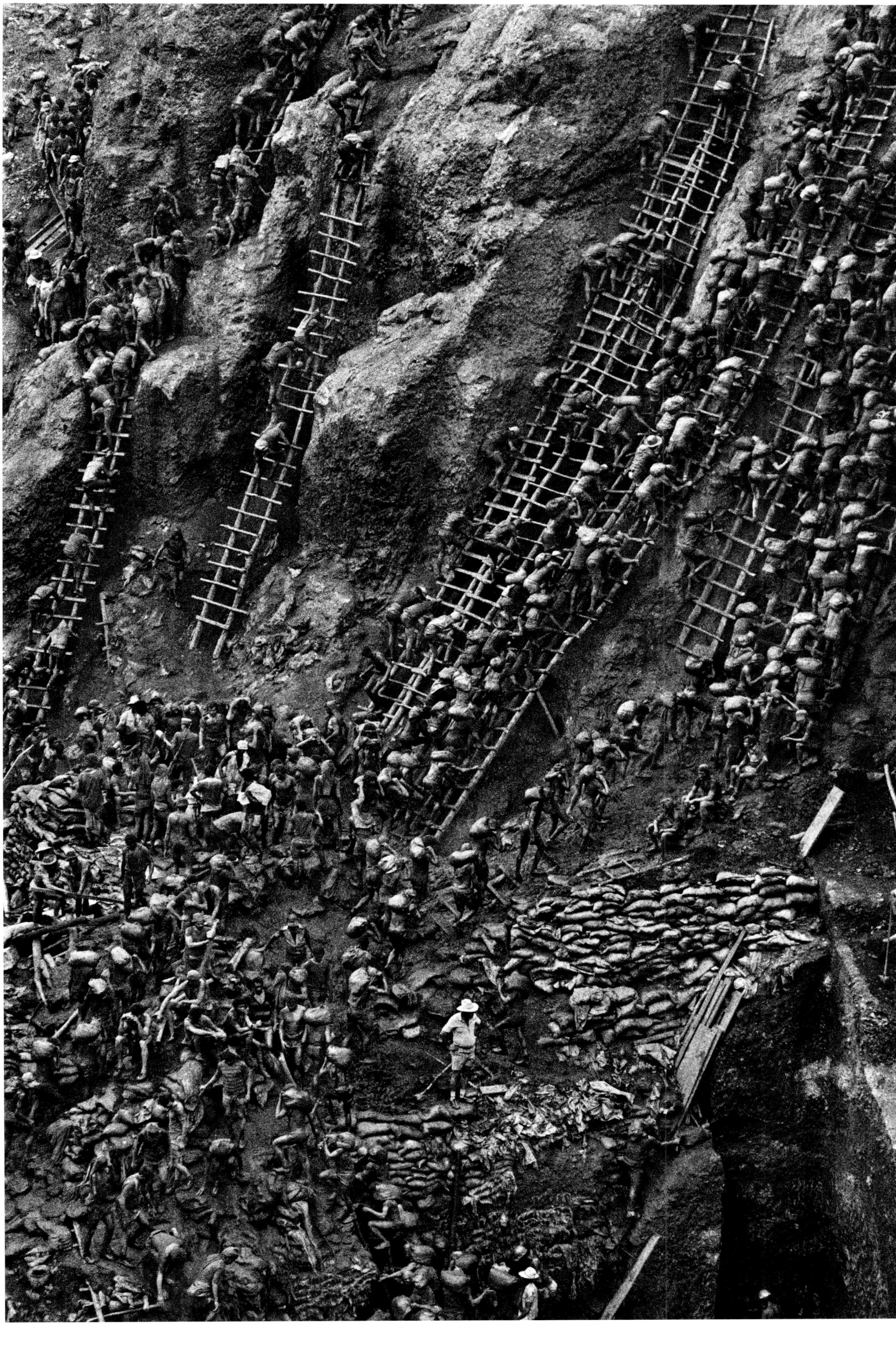

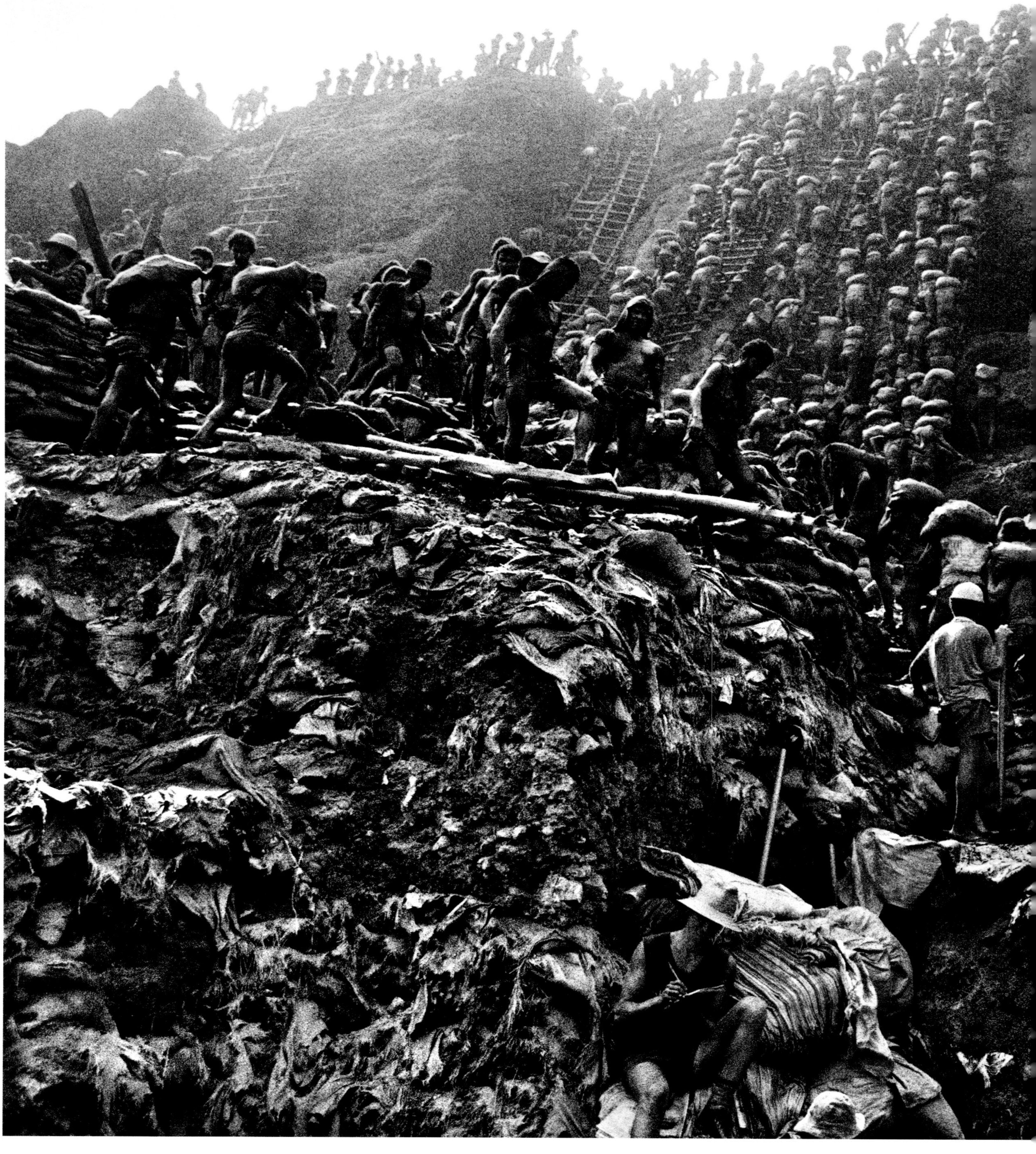

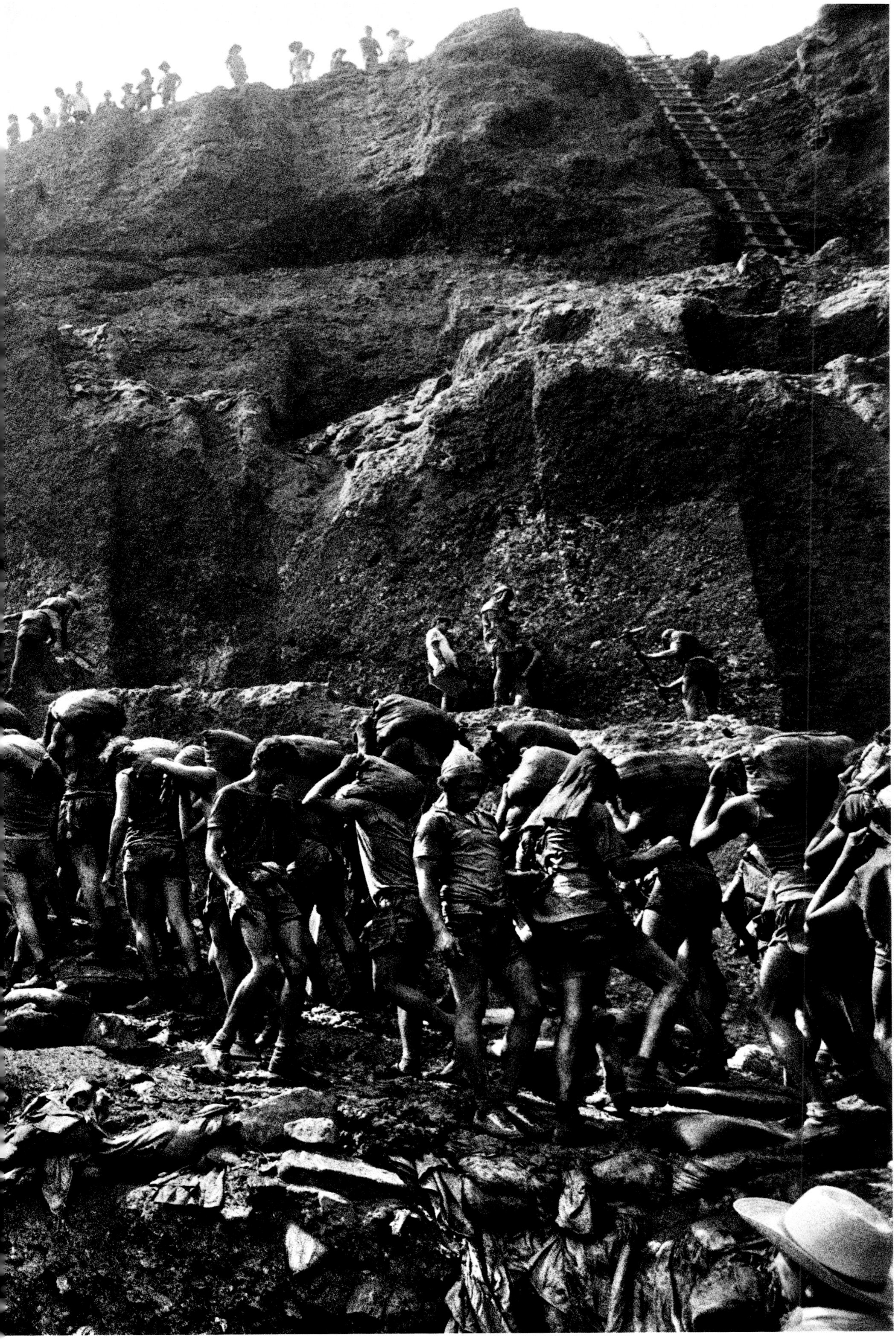

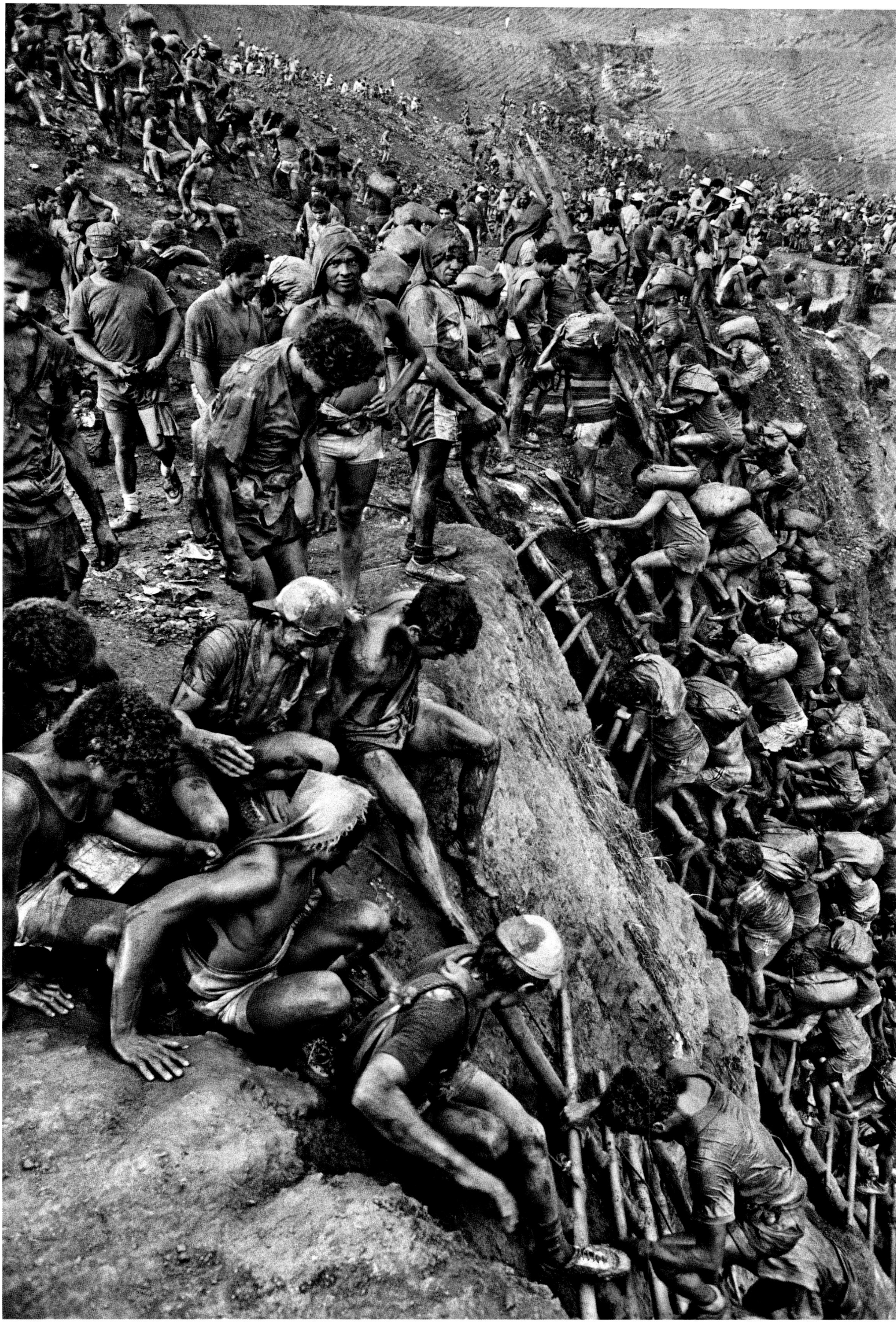

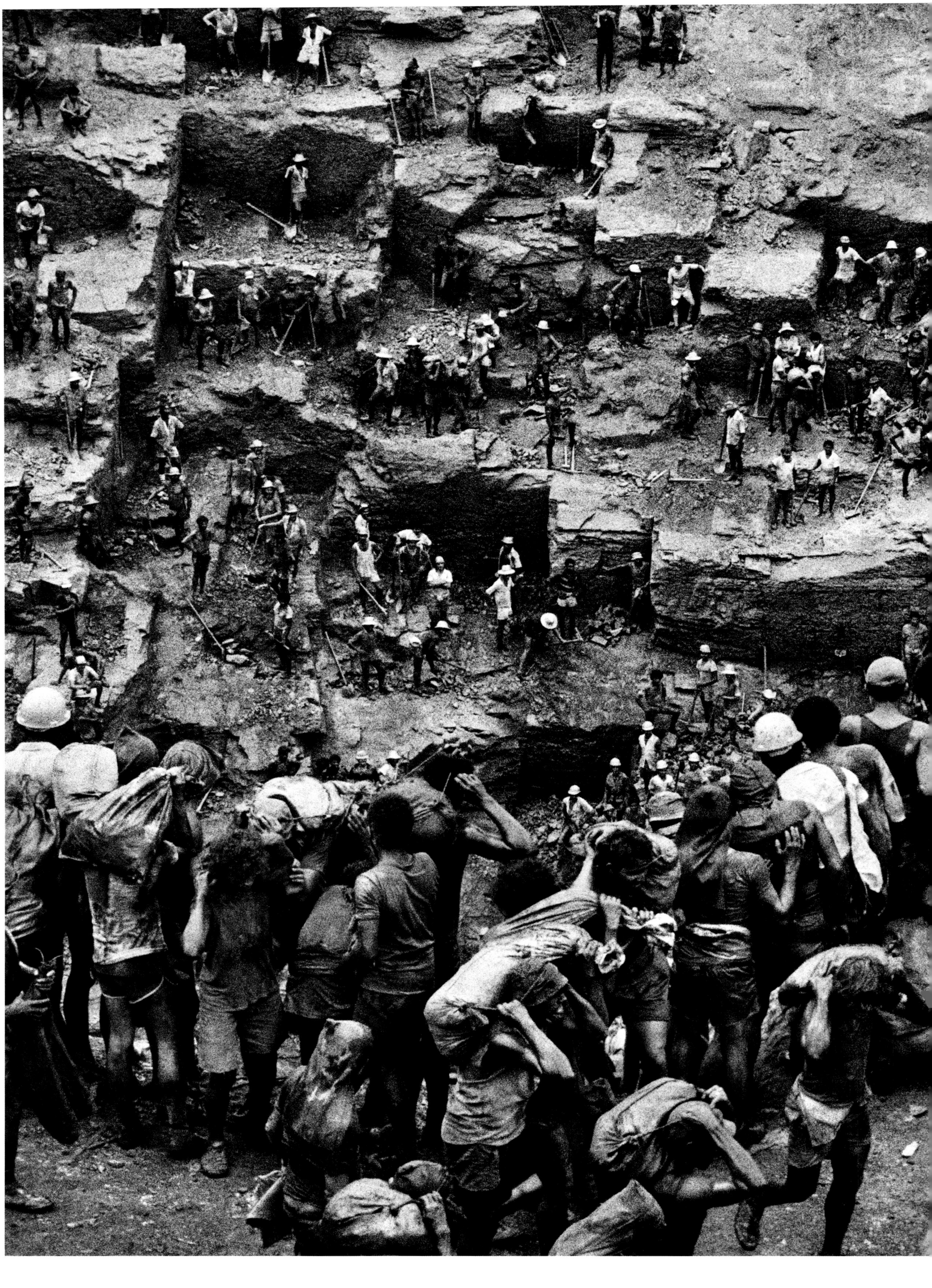

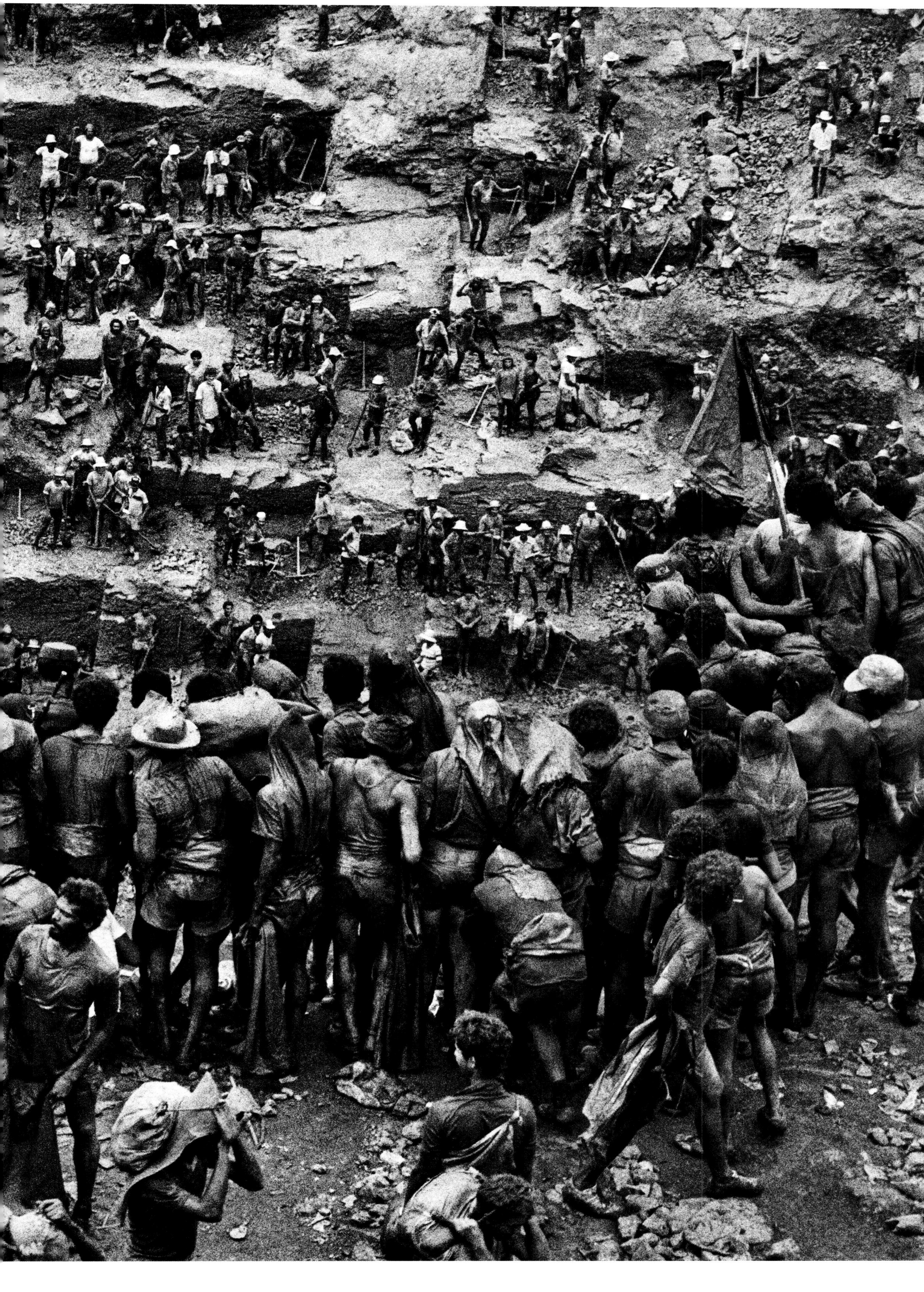

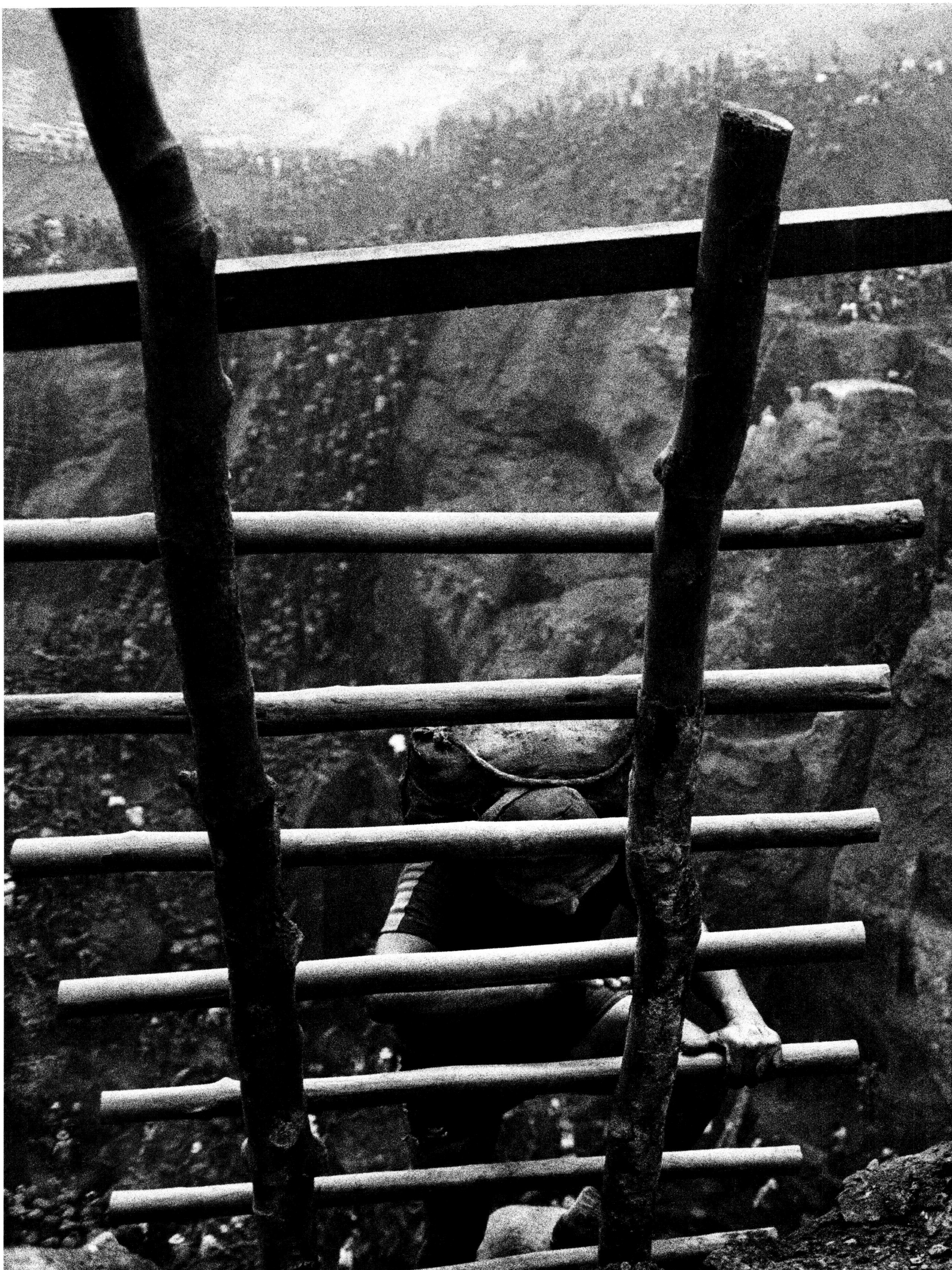

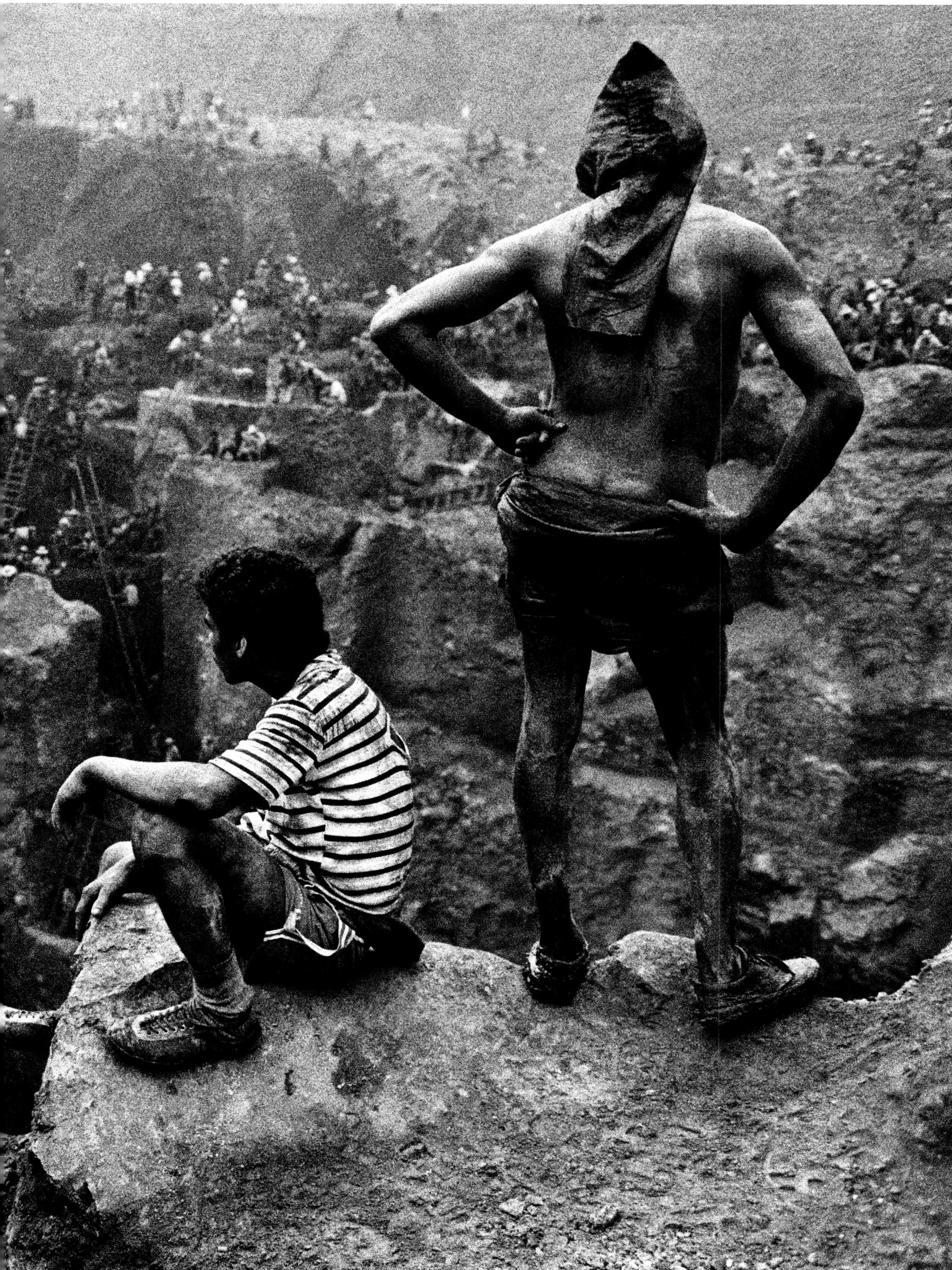

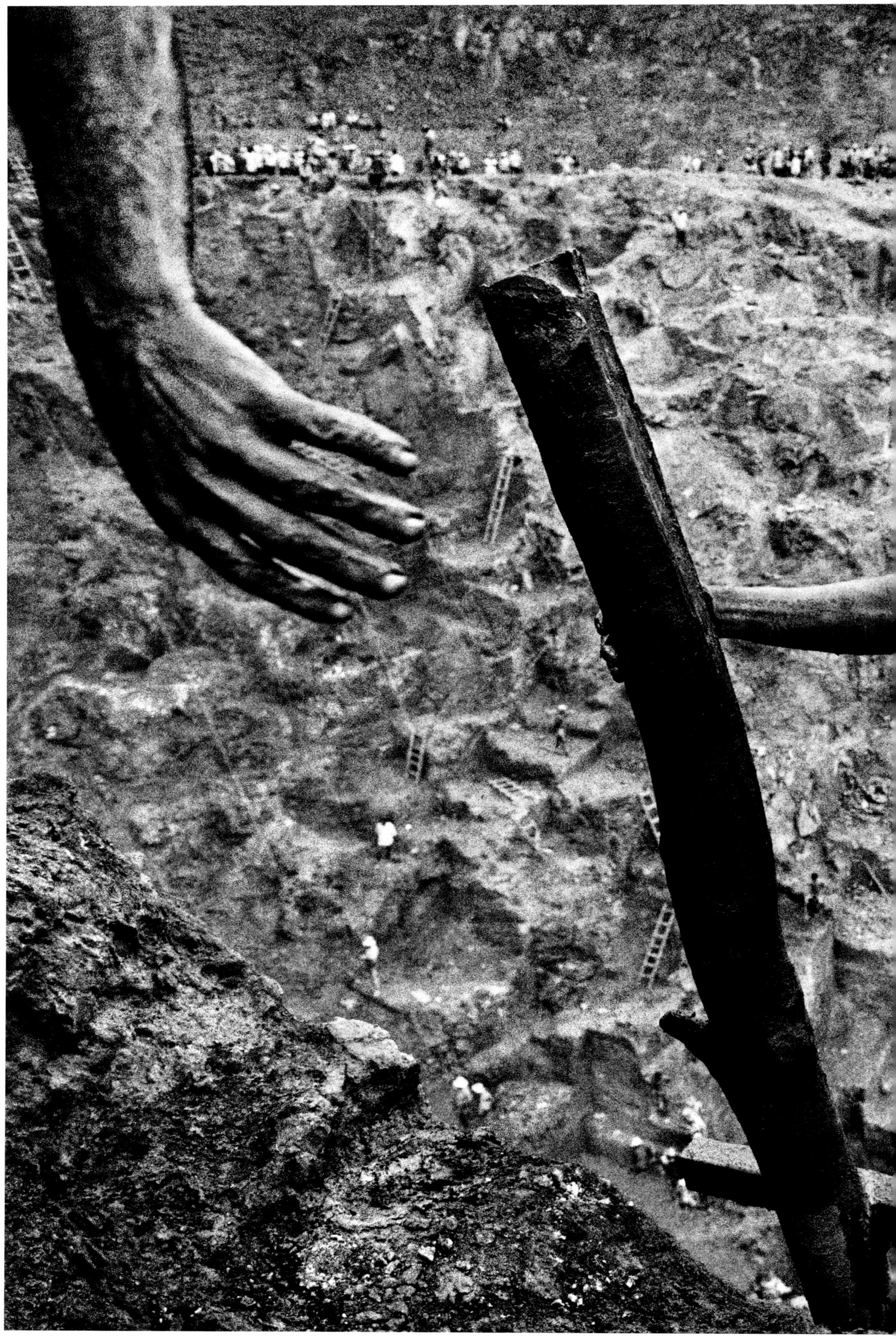

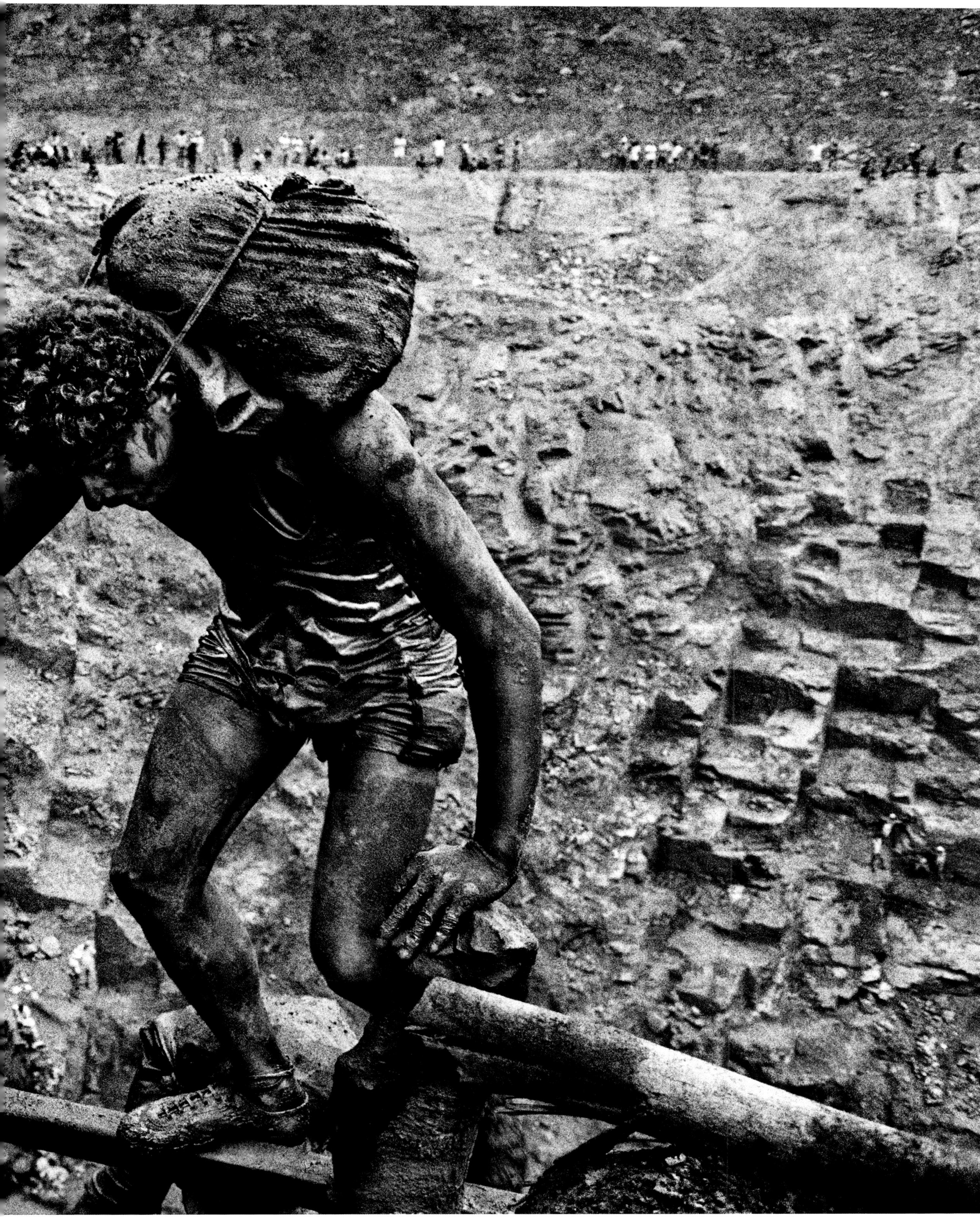

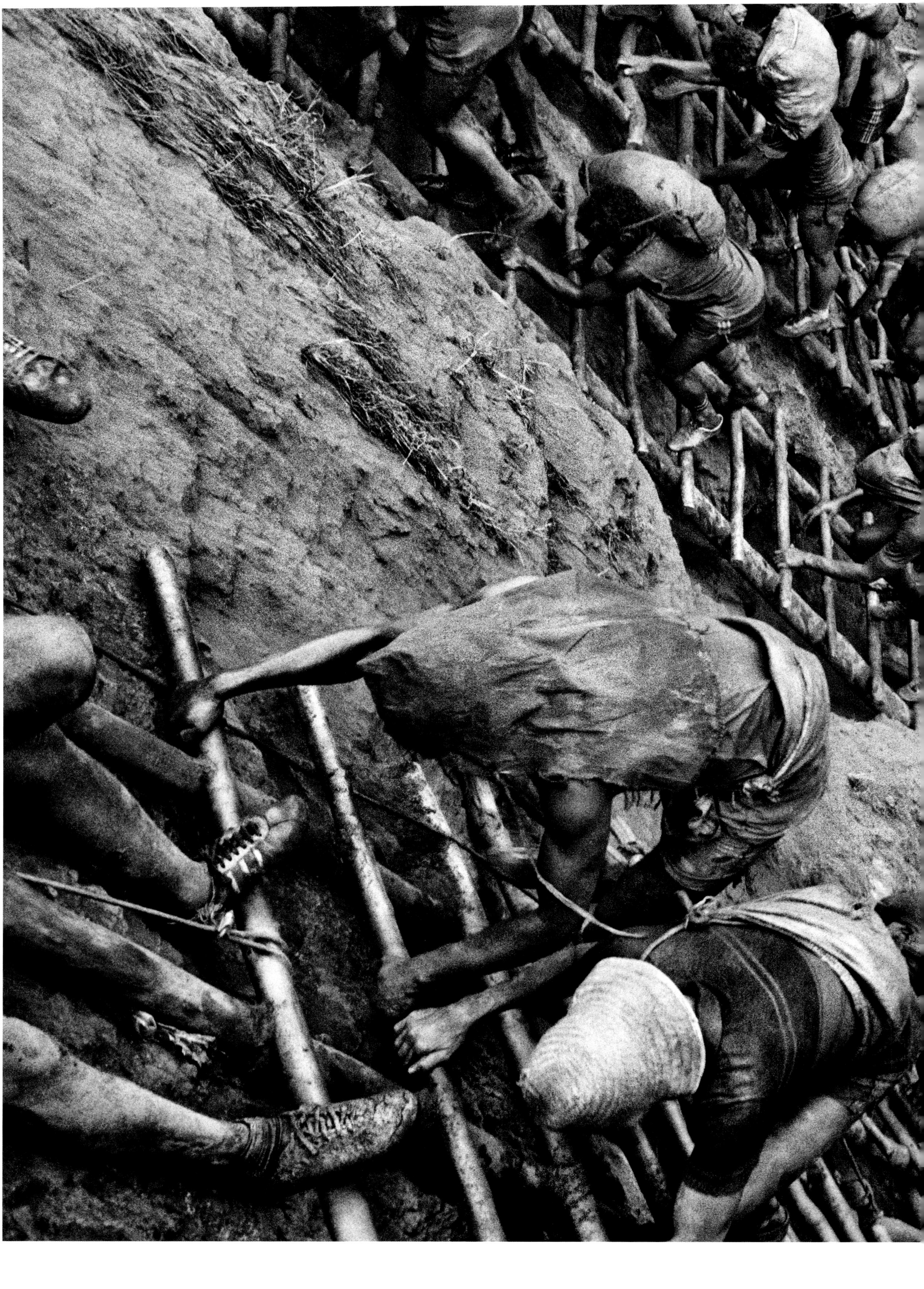

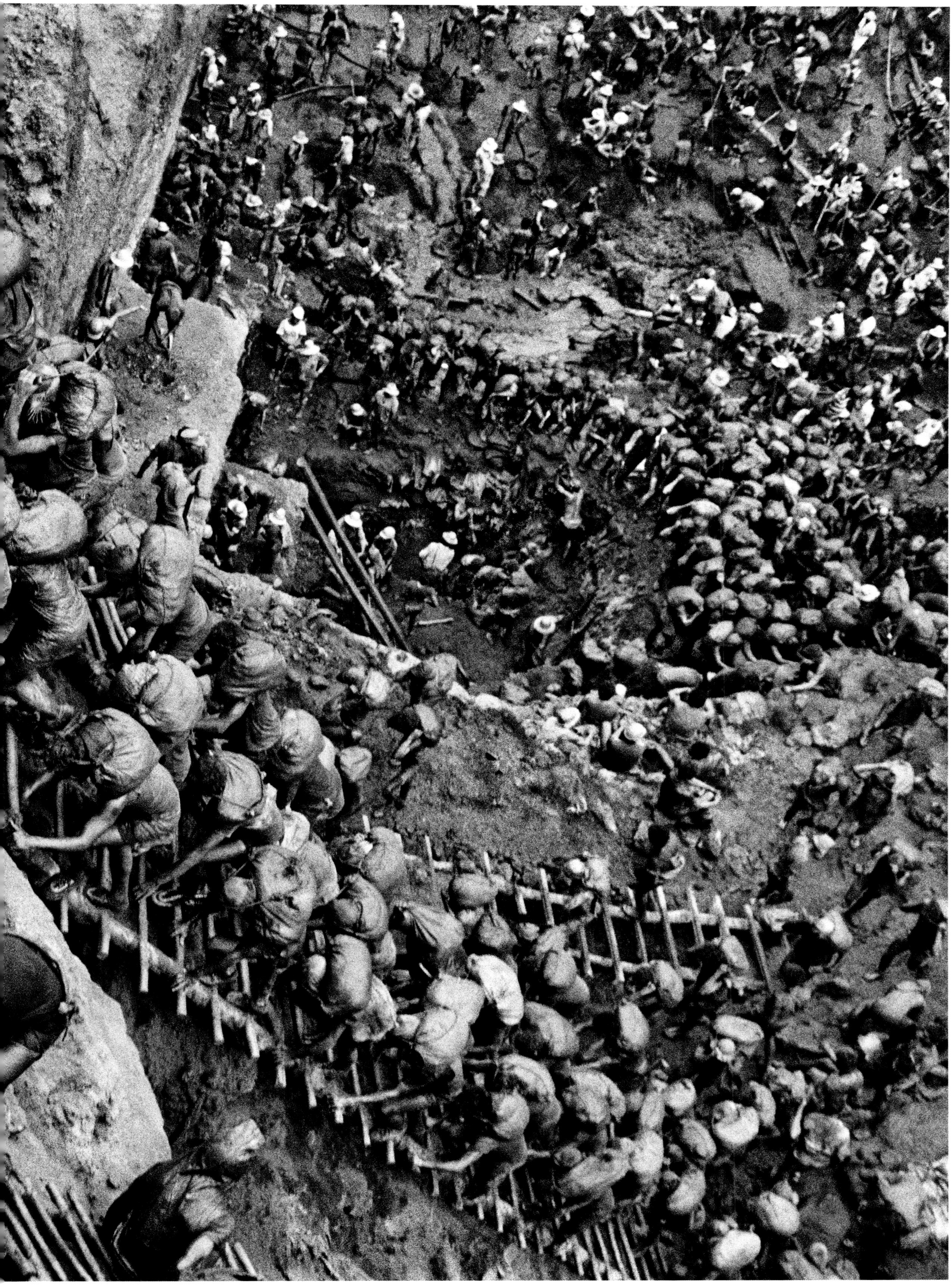

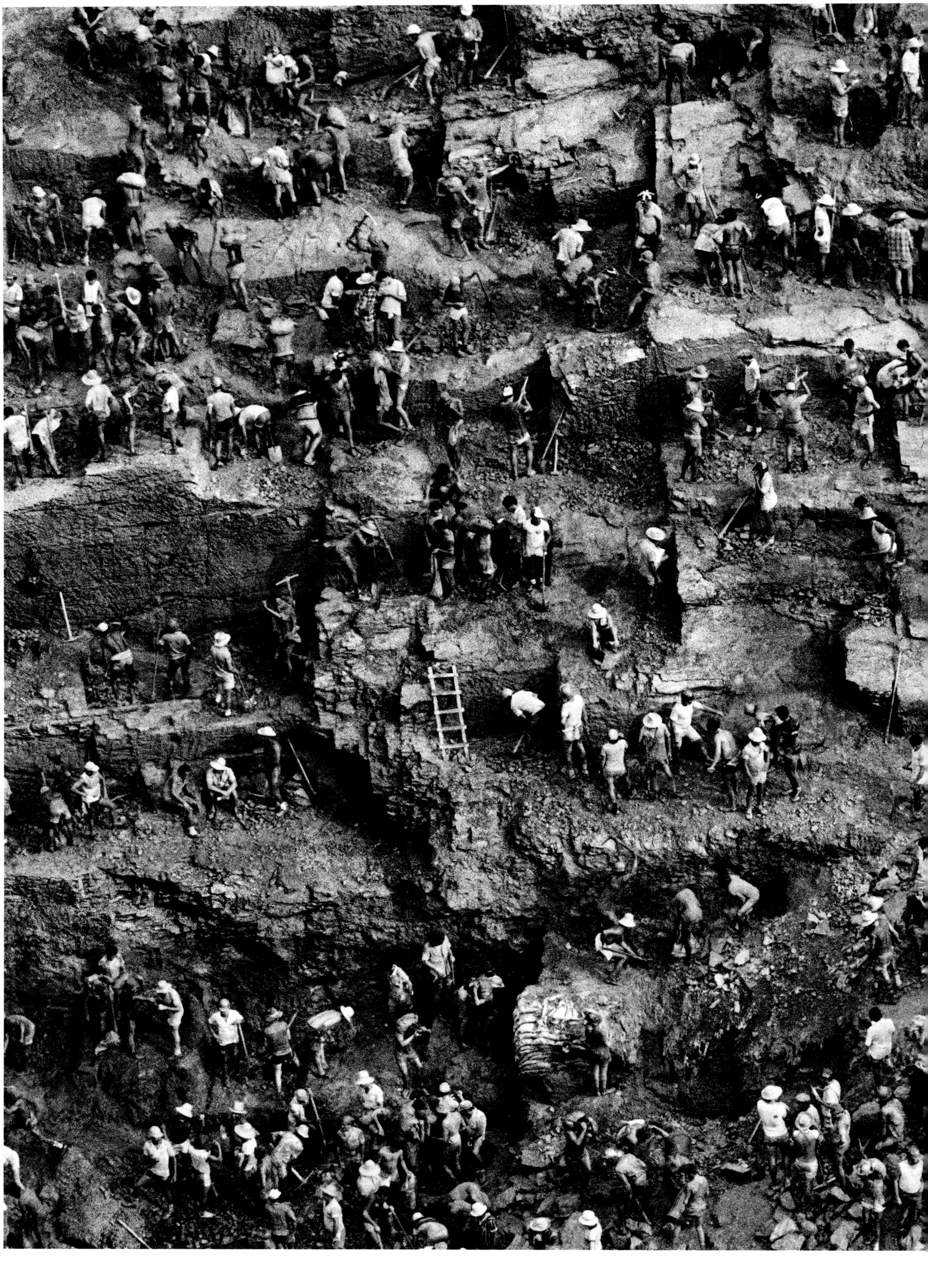

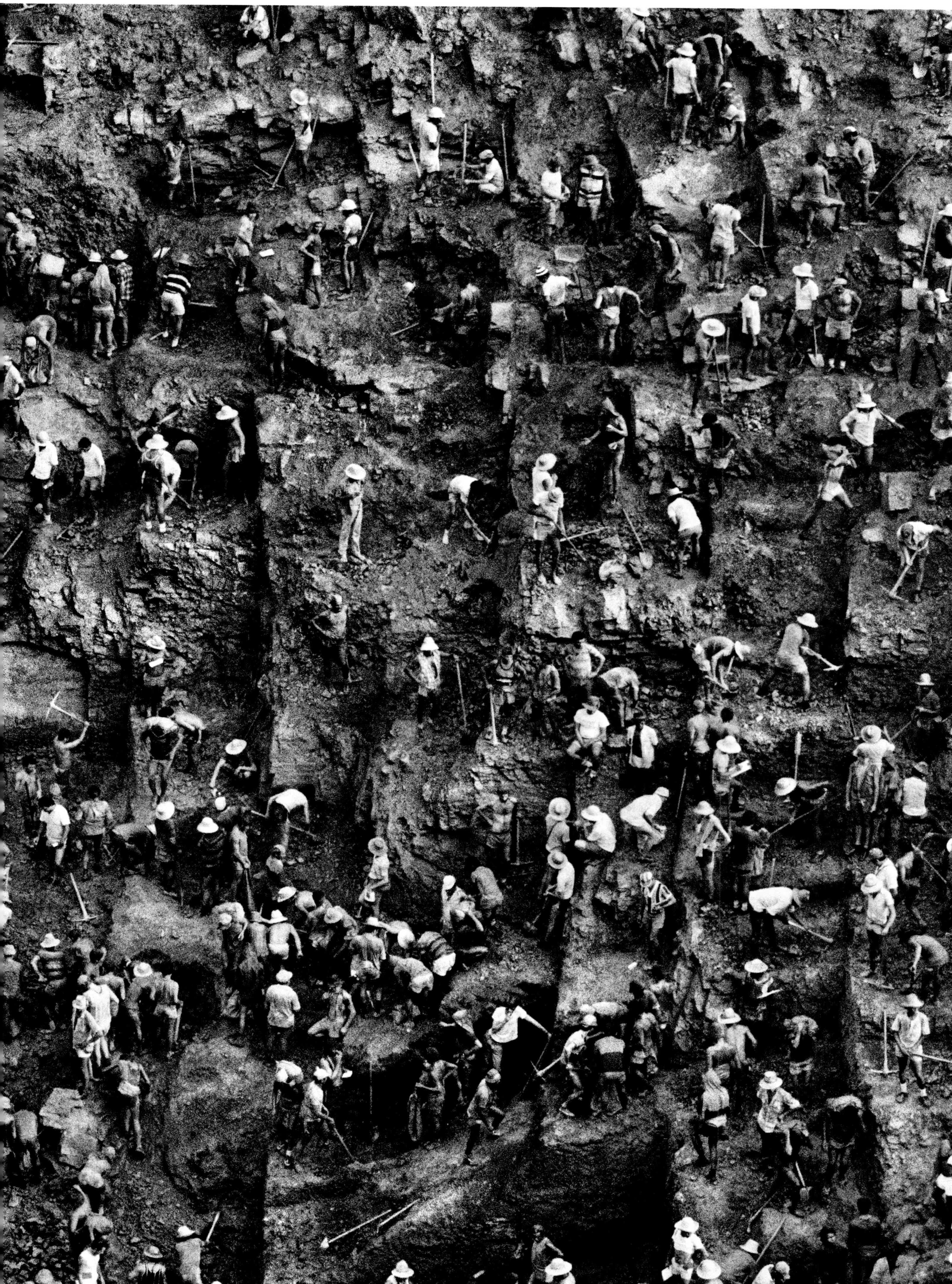

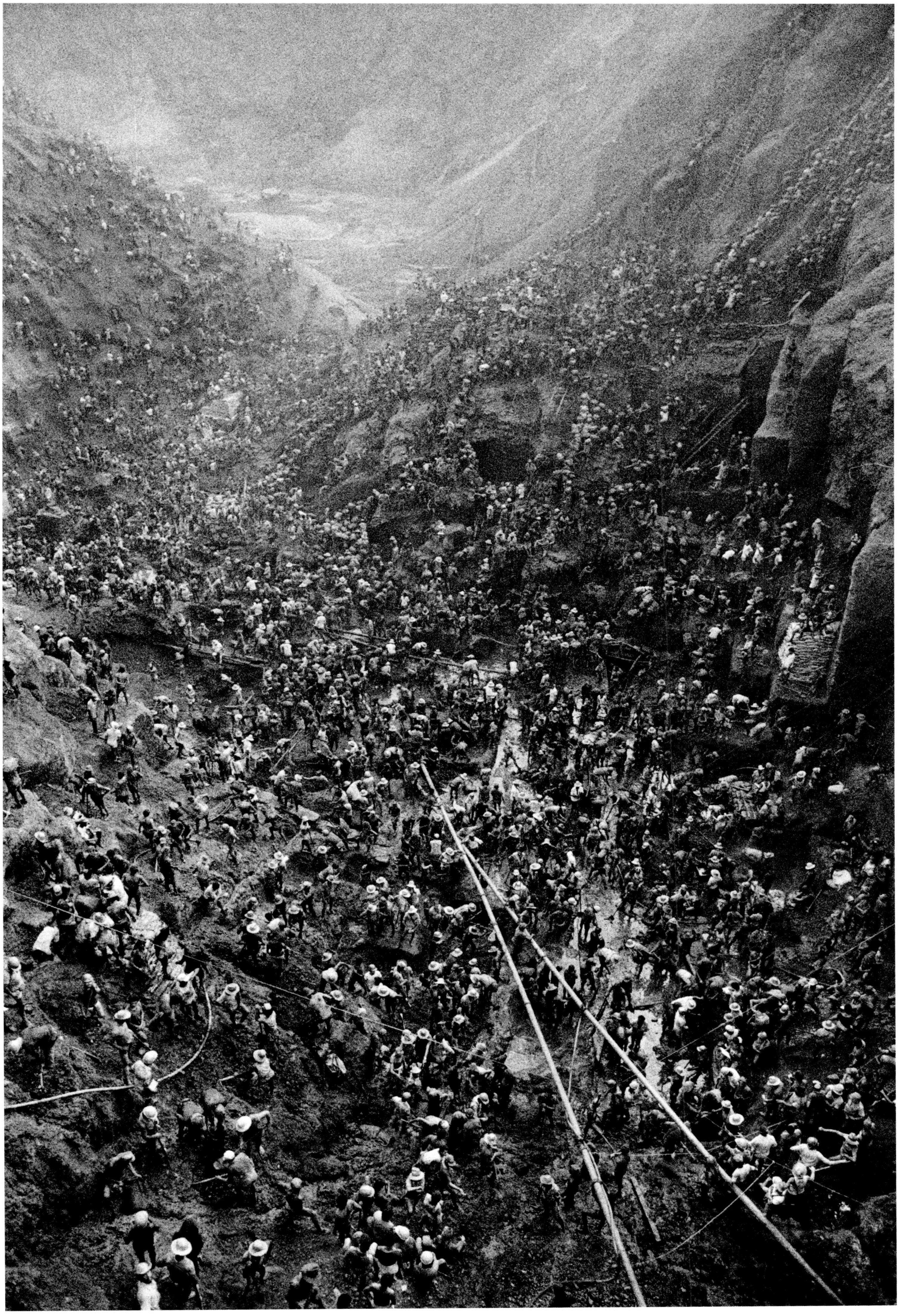

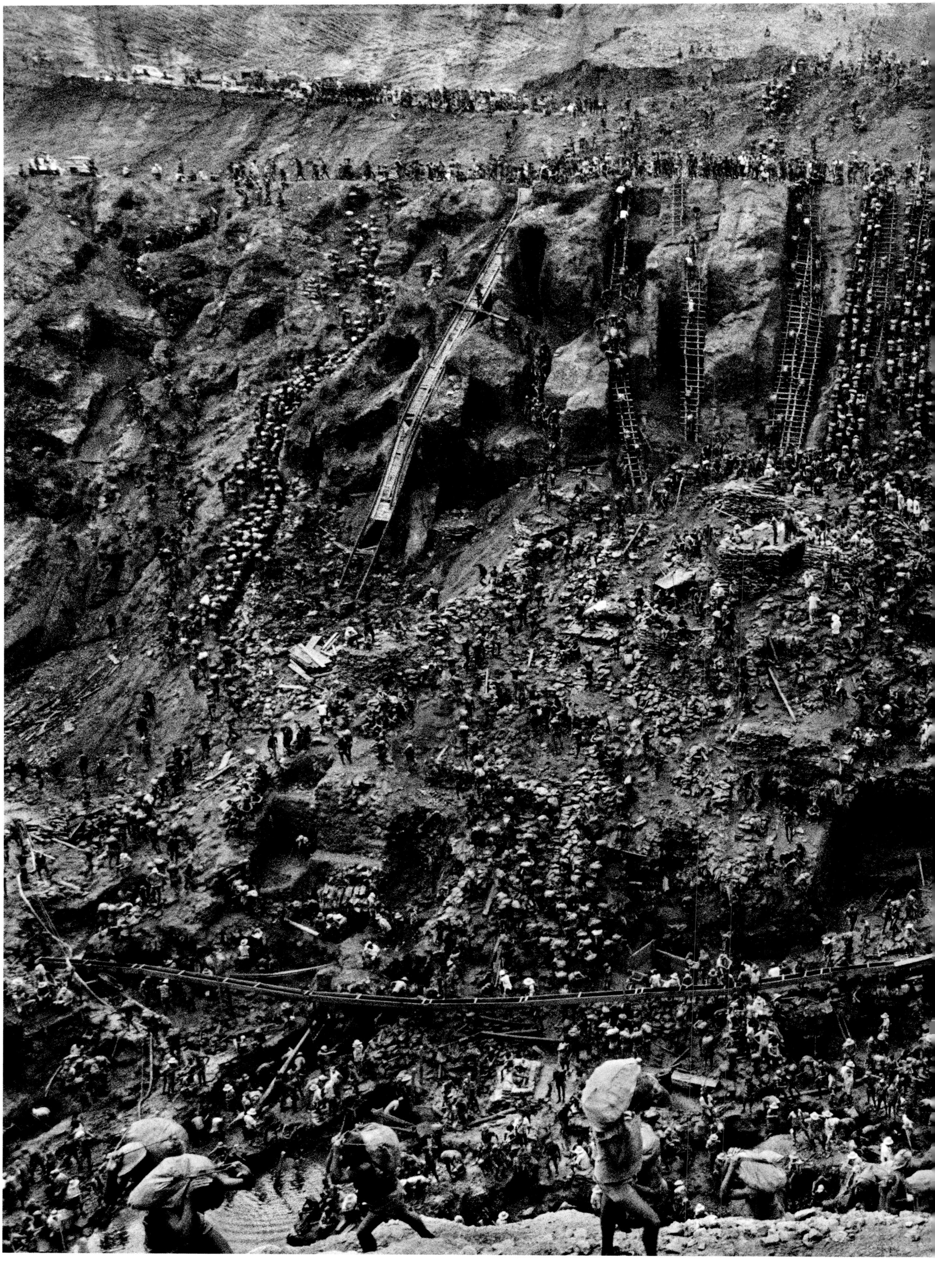

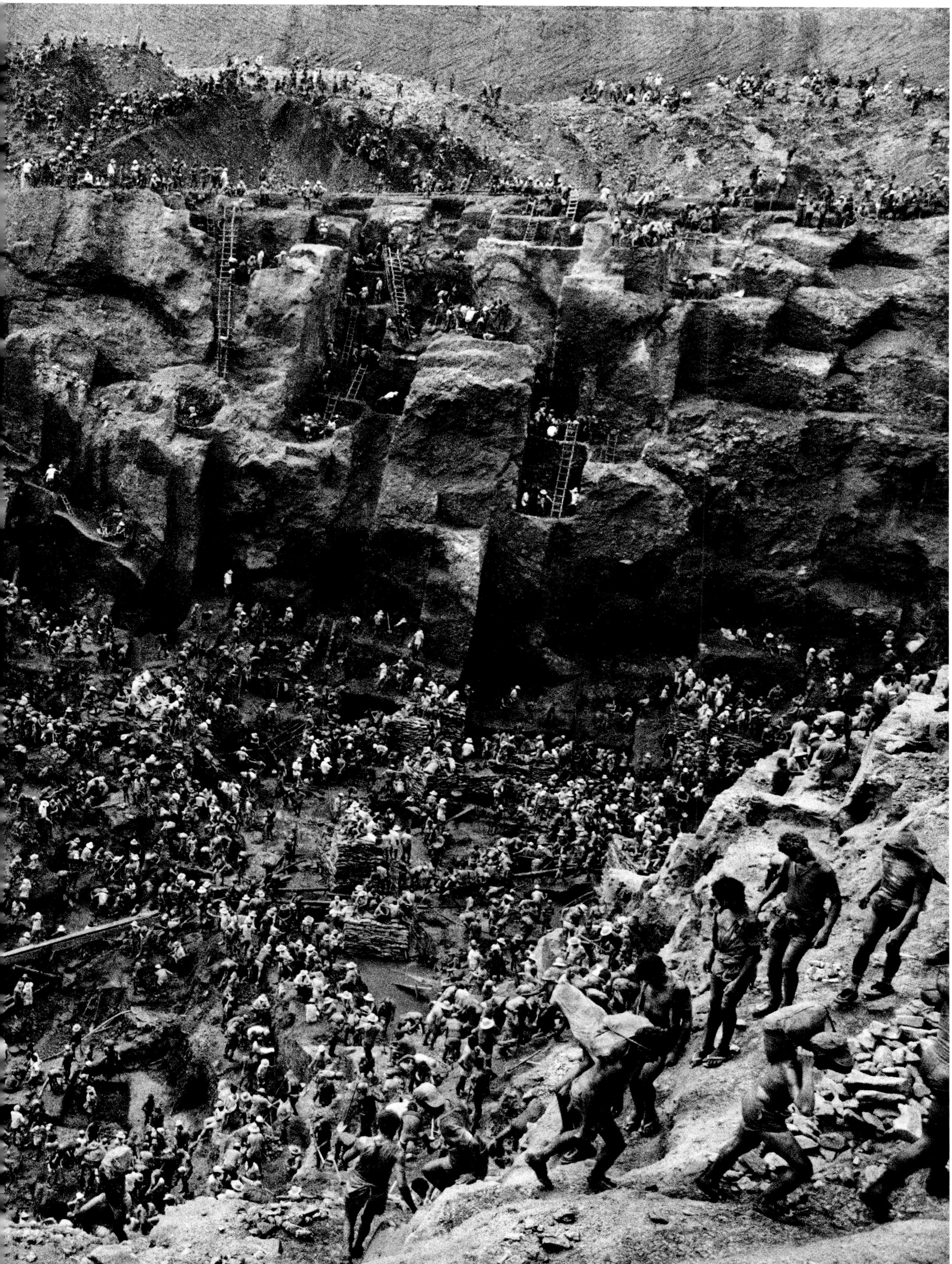

SERRA PELADA, THE RETURN OF BLACK AND WHITE

When Sebastião Salgado was finally authorized to visit Serra Pelada in September 1986, having been blocked for six years by Brazil's military authorities, many journalists, photographers, and filmmakers had already reported extensively on the gold mine and its teeming mass of workers. Salgado, however, was undeterred. His aim was not to make news with his images. Rather, it was to include them in the long-term photographic project that he was just beginning and would eventually be published and exhibited as *Workers: An Archeology of the Industrial Age*.

Why, then, did Salgado's photographs become the iconic images of Serra Pelada?

Discounting talent, patience, and courage, what most distinguished them from earlier coverage was that they were in black and white. Salgado had previously shot many magazine stories in color, including one we did together about Mexico's oil boom for *The New York Times Magazine* in 1981. But while he "lived off color," as he put it, he had always preferred "the *chiaroscuro* palette of black-and-white images." His first book, *Other Americas*, a poetic portrait of the backlands of Latin America, told its story in black and white. And for *Workers* and later books, that was again his choice.

In this sense, Salgado saw himself following a tradition whose masters, from Edward Weston and Brassaï to Robert Capa and Henri Cartier-Bresson, had dominated the early and mid-20th century. But from a commercial point of view, this was a risky path. By the 1970s, black-and-white photography had come to seem outdated. Although daily newspapers had yet to introduce color to their pages, color images were overwhelmingly favored by editors of the likes of *Life*, *Time*, *Newsweek*, *Paris Match*, and the Sunday magazines of major British and American newspapers.

The reason was simple: Technological advances in printing had by then made it possible for news magazines to publish reportages using high-resolution color photography. And the main appeal of this was that they could charge significantly more for advertisements in color than in black and white. For senior editors, then, it made little sense to publish photojournalistic essays in black and white in what were known as *color* supplements and magazines.

Salgado's *Serra Pelada* portfolio would prove an exception. It also marked a return to grace of black-and-white photography.

It happened somewhat accidentally. When Salgado reached home in Paris in early autumn 1986, he showed a selection of images from the mine to Jimmy Fox, editor of the renowned Magnum photo agency, of which Salgado was a member. Impressed, Fox suggested they be syndicated to the agency's regular clients, but Salgado was in no hurry. He was busy preparing his next journey for *Workers* and it was only the following spring that he sent *Serra Pelada* to Neil Burgess at Magnum's new office in London. He thought it might interest the literary magazine, *Granta*.

"My experience at the time was that if I proposed a story to one of the color magazines and said the photographer wanted to shoot it in black and white, the immediate answer was no," Burgess recalled. "But when I saw Sebastião's images, I said, 'I'm not giving them to *Granta*.' Instead I took them straight round to Michael Rand, art director of the *Sunday Times Magazine,* London. He asked me how much, I doubled what I'd planned to ask and he said yes."

On May 24, 1987, the newspaper's color magazine published six pages of *Serra Pelada* in black and white and the reaction was immediate. Other magazine editors came knocking on Burgess's door. The Victoria & Albert Museum's director of photography wanted to buy some prints for its collection. As did Paul Arden, creative director of the powerful advertising agency Saatchi & Saatchi.

"I told Paul Arden, 'Why not offer him a job?'" Burgess said, aware that Salgado needed funds to continue his global travels for *Workers*. And as a result, Salgado did two shoots for the agency.

Around the same time, a package of *Serra Pelada* photographs reached Robert Dannin, Magnum's editor in the United States. He immediately took them to *Life*, only to be told that the magazine had already run color pictures of the mine and was not interested. So Dannin's next stop was Peter Howe, the photo editor at *The New York Times Magazine*.

"I remember Peter saying, 'I'll get six to eight pages—or I'm leaving this job,'" Dannin said.

More than three decades later, Howe also recalls his reaction. "It blew me away, an amazing piece of work, biblically epic," he said of the *Serra Pelada* portfolio. "I took it in to the senior editors and laid out the photos. And something extraordinary happened. There was complete silence—the sort of silence that is working for you, not against. In my entire career at *The New York Times*, I never saw editors react to any set of pictures as they did to *Serra Pelada*."

To his frustration, Howe did not get the cover of the July 7, 1987, edition of the magazine, but it carried eight inside pages on *Serra Pelada* and, as in London, the impact was immediate. "Most of the feedback came from the photo community," he said, "because it was not normal for any photographer to get that much space in the magazine in black and white. It gave Sebastião a prestige in the United States that up to that point I don't think he really had."

As the man charged with marketing Salgado's photos, Dannin sensed an even stronger response. "It revolutionized the industry," he said of the *Serra Pelada* story, "not only in advertising but also in books. Even the U.S. Army began running recruiting ads in black and white. There was a lot of imitation."

Starting with *Life* magazine, Dannin was now also able to sell a syndicated package of Salgado's photographs to magazines across the world, with the understanding that they would publish "chapters" of *Workers* as they came in—or pay a kill fee. And this embraced many new outlets, like *Paris Match* and *Libération* in France, *Stern* in Germany, *Il Venerdì di Repubblica* in Italy, *El País* in Spain and many more.

In that same year, 1987, Salgado shot his last magazine story in color when he covered the 70th anniversary of the Soviet Revolution. Since then, along with *Workers* and two subsequent long-term projects, *Migrations* and *Genesis*, all his photographs have appeared in black and white, as much in exhibitions and books as in the magazines. And they have gone from illustrating stories to stand-alone reportages to becoming valued collectors' items.

What, then, is the distinct appeal of black-and-white photography?

"After *Serra Pelada* people woke up to the value of black-and-white photography," Neil Burgess said. "Not just as a counterpoint to full color, but as a different means of expression, a different form of photography. A photographer can control and develop the narrative more clearly and directly in black and white, while color can distract; it's a different aesthetic."

For Peter Howe, black and white added to the force of *Serra Pelada*. "Part of its power was that the whole environment seemed so foreign to people in the U.S. in the late 20th century," he said. "It had an almost medieval quality, as if it were surprising to find people like the miners still living like that. Black and white also gave them a quality that reinforced the idea that this was as much art as it was journalism."

What most struck Robert Dannin was the "cinematic quality" of the *Serra Pelada* images. "It makes you think of black-and-white films like those of Eisenstein or Kurosawa," he said. "I'm not saying Sebastião is a movie director, but there's a choreography, an orchestration, to the way it is shot. In *Workers* too, he seems to be looking at the industrial age through a rear-view mirror—and in black and white, which is how we experienced it."

For Salgado, his preference for black and white was never in question. From his childhood, when his fair skin forced him to keep out of Brazil's fierce sun, he became fascinated by the play of light and shadows. Two decades later, when he began developing his technique as a photographer, he saw how this contrast served to heighten reality.

In contrast, he often felt distracted by color. "When I looked through the lens, I knew that those colors would be more important than the people I photographed. Black and white provides an abstraction because in its tones and shades you can sense the dignity and power of people."

Serra Pelada illustrates this. While color photographs of the miners are dominated by the ochre dust covering them from head to foot, in black and white every detail leaps out: their exhausted faces, their muscular arms and legs, their damp and dirty clothes. Again, in panoramic shots of the cavernous mine, black and white allows points of light to identify every worker as he climbs the ladder with his sack full of earth.

Today, with photography increasingly absorbed by the art world and subject to digital manipulation, the vintage appearance of black-and-white images no doubt contributes to their appeal. Yet, no less than those of past masters, Salgado's black-and-white photographs project an immediacy that makes them vividly contemporary. We know that the mine at Serra Pelada is now closed, yet the intense drama of the gold rush leaps out of these images. Somehow, in black and white, time has a way of standing still.

Alan Riding
2019

SERRA PELADA, DIE RÜCKKEHR VON SCHWARZ UND WEISS

Als Sebastião Salgado im September 1986 endlich die Erlaubnis erhielt, Serra Pelada zu besuchen, nachdem er zuvor von der brasilianischen Militärregierung sechs Jahre lang daran gehindert worden war, hatten bereits viele Journalisten, Fotografen und Filmemacher ausführlich über die Goldmine und die zahllosen dort arbeitenden Männer berichtet. Aber Salgado ließ sich davon nicht abhalten. Sein Ziel war es nicht, mit seinen Bildern Schlagzeilen zu machen. Vielmehr wollte er sie in ein langfristig angelegtes Fotoprojekt integrieren, mit dem er gerade erst begonnen hatte und das schließlich unter dem Titel *Workers: An Archeology of The Industrial Age* (*Arbeiter: Zur Archäologie des Industriezeitalters*) publiziert und ausgestellt werden sollte.

Aber warum sind Salgados Aufnahmen von Serra Pelada zu ikonischen Bildern geworden?

Lässt man Talent, Geduld und Mut einmal beiseite, unterscheiden sie sich von früheren Reportagen vor allem dadurch, dass es Schwarz-Weiß-Aufnahmen sind. Zuvor hatte Salgado für Zeitschriften viele Fotostrecken in Farbe aufgenommen, darunter eine, die ich 1981 mit ihm zusammen für das *New York Times Magazine* über den mexikanischen Ölboom produziert hatte. Aber während er „von der Farbe lebte", wie er es ausdrückte, hat er persönlich immer „das Chiaroscuro der Schwarz-Weiß-Bilder" bevorzugt. Sein erstes Buch, *Other Americas*, ist ein poetisches Porträt des lateinamerikanischen Hinterlandes, das seine Geschichte in Schwarz und Weiß erzählt. Und für *Workers* und spätere Publikationen wählte er erneut Schwarz-Weiß.

In dieser Hinsicht folgte Salgado einer fotografischen Tradition, deren Meister – von Edward Weston und Brassaï bis hin zu Robert Capa und Henri Cartier-Bresson – das frühe und mittlere 20. Jahrhundert dominiert hatten. Kommerziell war dies jedoch ein riskanter Weg. In den 1970er-Jahren schien sich die Schwarz-Weiß-Fotografie überlebt zu haben. Obwohl Tageszeitungen in jener Zeit noch nicht in Farbe gedruckt wurden, bevorzugten Zeitschriften wie *Life*, *Time*, *Newsweek*, *Paris Match* und die Sonntagsbeilagen der großen britischen und amerikanischen Tageszeitungen Farbaufnahmen.

Der Grund war einfach: Der technische Fortschritt beim Drucken ermöglichte es den Zeitschriften, für ihre Reportagen hochauflösende Farbfotografien zu verwenden. Und der große

Reiz lag darin, dass sie nun für Werbung in Farbe weitaus mehr berechnen konnten als für Werbung in Schwarz-Weiß. Für die Chefredakteure machte es daher wenig Sinn, in den *Farb*-Beilagen und Magazinen fotojournalistische Essays in Schwarz-Weiß zu publizieren.

Salgados *Serra Pelada*-Portfolio war hier eine Ausnahme. Es markierte zudem eine Rückkehr zur Eleganz der Schwarz-Weiß-Fotografie.

Das Ganze passierte eher zufällig. Als Salgado im Frühherbst 1986 wieder in seiner Wahlheimat Paris eintraf, zeigte er Jimmy Fox, dem Redakteur der renommierten Fotoagentur Magnum, deren Mitglied Salgado damals war, eine Auswahl seiner Bilder der Goldmine. Tief beeindruckt schlug Fox vor, sie den Stammkunden der Agentur anzubieten, aber Salgado hatte es nicht eilig. Er war dabei, seine nächste Reise für das Projekt *Workers* vorzubereiten, und erst im folgenden Frühjahr schickte er das *Serra Pelada*-Portfolio an Neil Burgess in das neue Londoner Büro von Magnum. Er dachte, es könnte vielleicht für das Literaturmagazin *Granta* interessant sein.

„Meine damalige Erfahrung war, dass, wenn ich einem Farbmagazin eine Geschichte mit dem Hinweis anbot, der Fotograf plane sie in Schwarz-Weiß, die direkte Antwort ein Nein war", erinnerte sich Burgess. „Aber als ich Sebastiãos Bilder sah, sagte ich mir, ich werde sie nicht *Granta* anbieten. Stattdessen brachte ich sie direkt nach nebenan zu Michael Rand, dem Art Director des Londoner *Sunday Times Magazine*. Er fragte mich, was sie kosten sollten, und ich verdoppelte spontan den Preis, den ich im Sinn hatte, und er stimmte zu."

Am 24. Mai 1987 veröffentlichte das Farbmagazin der Zeitung *Serra Pelada* auf sechs Seiten in Schwarz-Weiß, und die Reaktion folgte umgehend. Weitere Zeitschriftenredakteure klopften bei Burgess an. Der Leiter der fotografischen Sammlung des Victoria & Albert Museum wollte einige Abzüge für die Sammlung kaufen – ebenso Paul Arden, der Kreativdirektor der einflussreichen Werbeagentur Saatchi & Saatchi.

„Ich fragte Paul Arden, ob er Salgado nicht einen Auftrag anbieten wolle", sagte Burgess. Er wusste, dass Salgado Geld brauchte, um seine langen Reisen rund um den Globus für *Workers* zu finanzieren. Das Ergebnis war, dass Salgado zweimal für die Agentur fotografierte.

Ungefähr zur gleichen Zeit erreichte ein Pak von *Serra Pelada*-Fotografien Robert Dannin, Bildredakteur bei Magnum in den Vereinigten Staaten. Er bot sie sofort *Life* an, aber man erklärte ihm dort, die Zeitschrift habe schon Farbaufnahmen der Mine veröffentlicht und man sei nicht interessiert. Daraufhin wandte er sich an Peter Howe, den Bildredakteur des *New York Times Magazine*.

„Ich erinnere mich, wie Peter sagte: ‚Ich will sechs bis acht Seiten mit diesen Bildern, oder ich kündige'", berichtete Dannin.

Mehr als drei Jahrzehnte später erinnert sich auch Howe an seine Reaktion. „Es haute mich einfach um, es war ein ganz außergewöhnliches Werk, ein biblisches Epos", so charakterisiert er das *Serra Pelada*-Portfolio. „Ich trug die Fotos zu den Chefredakteuren und breitete sie vor ihnen aus. Und etwas Außergewöhnliches passierte. Es herrschte völlige Stille – jene Art von Stille, die für einen, nicht gegen einen ist. Während meiner gesamten Karriere bei der *New York Times* habe ich nie erlebt, dass Redakteure auf Bilder so reagierten wie auf die *Serra Pelada*-Serie."

Zu seiner großen Enttäuschung bekam Howe nicht das Cover der Ausgabe vom 7. Juli 1987, aber der *Serra Pelada*-Beitrag im Innenteil des Magazins umfasste acht Seiten. Wie in London war die Reaktion außergewöhnlich. „Das größte Echo kam aus der Fotogemeinde", sagte er, „weil es nicht üblich war, dass man einem Fotografen für Schwarz-Weiß-Bilder so viel Platz in einem Magazin einräumte. Das verschaffte Sebastião in den USA ein Ansehen, das er meiner Meinung nach bis zu diesem Zeitpunkt dort nicht gehabt hatte."

Dannin, der die Aufgabe hatte, Salgados Fotos zu vermarkten, erlebte eine noch stärkere Reaktion. „Die Serie revolutionierte die Branche", sagte er über die *Serra Pelada*-Bilder, „nicht allein in der Werbung, sondern auch bei den Büchern. Sogar die US-Armee suchte nun in Schwarz-Weiß-Anzeigen nach Rekruten. Es gab jede Menge Nachahmer."

Dannin schaffte es, ein Paket mit Salgados Fotografien zunächst an das Magazin *Life* und dann an weitere Zeitschriften in der ganzen Welt zu verkaufen mit der Übereinkunft, dass sie auch nachfolgende „Kapitel" aus *Workers* veröffentlichen würden, sobald sie vorlägen, oder aber ein Ausfallhonorar zahlten. Diese Vereinbarung wurde mit vielen neuen Abnehmern geschlossen, etwa *Paris Match* und *Libération* in Frankreich, dem *Stern* in Deutschland, *Il Venerdì di Repubblica* in Italien, *El País* in Spanien und vielen anderen.

Im selben Jahr, 1987, nahm Salgado seinen letzten Zeitschriftenbeitrag in Farbe auf, als er den 70. Jahrestag der Russischen Revolution fotografierte. Danach erschienen – zusammen mit *Workers* und zwei weiteren Langzeitprojekten, *Migrations* und *Genesis* – seine gesamten Fotografien nur noch in Schwarz-Weiß, in Ausstellungen und Büchern ebenso wie in Zeitschriften. Dabei haben sie sich von der Illustration von Geschichten über eigenständige Fotoreportagen zu geschätzten Sammlerstücken entwickelt.

Aber wo genau liegt der besondere Reiz der Schwarz-Weiß-Fotografie?

„Durch *Serra Pelada* wurden die Leute auf die Stärke der Schwarz-Weiß-Fotografie aufmerksam“, sagte Neil Burgess. „Nicht als Gegensatz zur Farbfotografie, sondern als ein anderes Ausdrucksmedium, eine andere Form der Fotografie. Ein Fotograf kann eine Erzählung in Schwarz-Weiß direkter und genauer kontrollieren und entwickeln als in Farbe, die eher vom Wesentlichen ablenkt. Es ist eine andere Ästhetik.“

Für Peter Howe hat das Schwarz-Weiß die Ausdruckskraft der *Serra Pelada*-Fotos verstärkt. „Ein Teil der Kraft dieser Bilder liegt darin, dass den Menschen in den USA im ausgehenden 20. Jahrhundert der Inhalt der Fotos so fremd war“, sagte er. „Sie hatten eine fast mittelalterliche Anmutung. Es war eine große Überraschung, dass Zeitgenossen noch wie diese Minenarbeiter leben. Und das Schwarz-Weiß gab den Bildern eine gleichermaßen künstlerische wie journalistische Qualität.“

Robert Dannin beeindruckte die „filmische Qualität“ der *Serra Pelada*-Aufnahmen am stärksten. „Das lässt an Schwarz-Weiß-Filme von Eisenstein und Kurosawa denken“, erklärte er. „Ich sage nicht, dass Sebastião ein Filmregisseur ist, aber seine Aufnahmen folgen einer Choreografie, einer Orchestrierung. Auch in *Workers* scheint er wie im Rückspiegel auf das Industriezeitalter zu schauen – in Schwarz und Weiß, so wie wir es erlebt haben.“

Salgado stellte seine Liebe zur Schwarz-Weiß-Fotografie nie infrage. Schon seit seiner Kindheit, als seine helle Haut ihn zwang, die unbarmherzige Sonne Brasiliens zu meiden, war er fasziniert vom Spiel von Licht und Schatten. Zwei Jahrzehnte später, als er seine fotografische Technik entwickelte, erkannte er, wie dieser Kontrast die Wirklichkeitswahrnehmung verstärkt.

Im Gegensatz dazu empfand er die Farbe oft als ablenkend. „Wenn ich durch das Objektiv sah, wusste ich, dass die Farben wichtiger sein würden als die Menschen, die ich fotografierte. Schwarz-Weiß sorgt für eine Abstraktion, denn in seinen Tönen und Schattierungen spürt man die Würde und Kraft der Menschen.“

Serra Pelada macht das deutlich. Während die Farbaufnahmen der Minenarbeiter von dem ockerfarbenen Staub beherrscht werden, der sie von Kopf bis Fuß bedeckt, springt in den Schwarz-Weiß-Aufnahmen jedes Detail scharf hervor: ihre erschöpften Gesichter, die muskulösen Arme und Beine, ihre feuchten, schmutzigen Kleider. Und in den Panoramabildern der riesigen Mine setzt das Schwarz-Weiß Lichtpunkte, die jeden einzelnen Arbeiter erkennen lassen, der mit seinem Sack voll Erde die Leiter hochsteigt.

Heute, da die Fotografie zunehmend von der Kunstwelt absorbiert und digital manipuliert wird, trägt der Vintage-Charakter der Schwarz-Weiß-Aufnahmen zweifellos zu ihrer

Attraktivität bei. Wie die Bilder der vergangener Meister haben Salgados Schwarz-Weiß-Fotografien eine Unmittelbarkeit, die wir als absolut zeitgenössisch erleben. Wir wissen, dass die Mine in Serra Pelada inzwischen geschlossen wurde, aber diese Bilder lassen das überwältigende Drama des Goldrauschs immer wieder aufleben. Es scheint, als stünde die Zeit in Schwarz-Weiß-Aufnahmen irgendwie still.

Alan Riding
2019

SERRA PELADA : LE RETOUR DU NOIR ET BLANC

Lorsqu'en septembre 1986 Sebastião Salgado reçut enfin l'autorisation de se rendre sur le site de Serra Pelada, qui lui avait été interdit pendant six ans par les autorités militaires brésiliennes, de nombreux journalistes, photographes et réalisateurs de film avaient déjà abondamment couvert cette mine d'or et la masse grouillante de ses prospecteurs. Il ne se sentit pas découragé pour autant. Il ne souhaitait pas faire du reportage d'actualité mais prendre des clichés pour son nouveau grand projet photographique, publié et exposé par la suite sous le titre de Workers: An Archeology of the Industrial Age – La Main de l'homme *en français.*

Dans ces conditions, pourquoi les photographies de Salgado sont-elles devenues les images les plus iconiques de Serra Pelada ?

Outre le talent, la patience et le courage de Salgado, ce qui distinguait ces clichés des précédents reportages, c'était le noir et blanc. Salgado avait réalisé de nombreux reportages en couleurs pour des magazines, dont un que nous avions fait ensemble sur le boom pétrolier au Mexique pour le New York Times Magazine *en 1981. Mais s'il « dépendait de la couleur pour travailler », comme il le disait, il avait toujours préféré « la palette en clair-obscur des images en noir et blanc ». Son premier livre,* Autres Amériques, *portrait poétique de régions méconnues d'Amérique latine, était déjà une narration en noir et blanc. Pour* La Main de l'homme *et les livres suivants, il fit à nouveau ce choix.*

En ce sens, Salgado s'inscrivait dans une tradition dont les maîtres, d'Edward Weston et Brassaï à Robert Capa et Henri Cartier-Bresson, avaient dominé le monde de la photographie du début au milieu du XXe siècle. D'un point de vue commercial, c'était cependant une voie risquée. Dans les années 1970, la photographie en noir et blanc était démodée. Même si les quotidiens n'avaient pas encore introduit la couleur dans leurs pages, celle-ci était généralement favorisée par les éditeurs de magazines comme Life, Time, Newsweek, Paris Match *et les suppléments dominicaux des grands journaux américains et britanniques.*

Et ce, pour une raison simple : les progrès technologiques accomplis dans le domaine de l'impression permettaient désormais aux magazines de publier des reportages photographiques avec des couleurs de haute résolution. Pour eux, l'attrait principal de cette technologie était de pouvoir facturer davantage les annonces publicitaires en couleurs. Les rédacteurs en chef avaient donc très peu d'intérêt à publier des reportages photo en noir et blanc dans ce que l'on appelait alors les « suppléments couleur » et les magazines.

Le portfolio Serra Pelada *de Salgado fut une exception qui marqua dans le même temps le retour en grâce de la photographie en noir et blanc.*

Tout arriva presque par accident. À son retour chez lui, à Paris, au début de l'automne 1986, Salgado montra une sélection de ses images de la mine à Jimmy Fox, rédacteur de la célèbre agence photographique Magnum, avec laquelle il était alors associé. Impressionné, Fox suggéra de proposer ces photos aux clients réguliers de l'agence, mais Salgado n'était pas pressé. Il préparait un nouveau voyage pour La Main de l'homme, *et ce n'est qu'au printemps suivant qu'il envoya les photos de* Serra Pelada *à Neil Burgess du nouveau bureau de Magnum à Londres, pensant qu'elles pourraient intéresser le magazine littéraire* Granta.

« D'après mon expérience, si je proposais un reportage à l'un des magazines couleur en disant que le photographe voulait faire du noir et blanc, on me répondait tout de suite "non", se rappelle Burgess. Mais lorsque j'ai vu les images de Sebastião, je me suis dit que je n'allais pas les donner à Granta. *Je suis allé les présenter directement à Michael Rand, directeur artistique du* Sunday Times Magazine *de Londres. Il m'a demandé combien, j'ai doublé ce que je pensais demander et il a dit oui. »*

Le 24 mai 1987, le Sunday Times Magazine *publiait ainsi* Serra Pelada *sur six pages en noir et blanc. La réaction fut immédiate. D'autres rédacteurs en chef de magazines vinrent frapper à la porte de Burgess, tandis que le directeur de la photographie du Victoria and Albert Museum voulut acheter des tirages pour sa collection, de même que Paul Arden, directeur de la création de la puissante agence de publicité Saatchi & Saatchi.*

« J'ai dit à Paul Arden : pourquoi ne pas lui passer une commande ? » raconte Burgess, sachant que Salgado avait besoin d'argent pour poursuivre ses voyages dans le monde entier pour La Main de l'homme. *L'agence signa ainsi deux contrats avec le photographe.*

Vers la même époque, les photos de Serra Pelada *arrivaient sur le bureau de Robert Dannin, directeur éditorial de Magnum aux États-Unis. Il les porta immédiatement chez* Life, *pour s'entendre dire que le magazine avait déjà publié un reportage en couleurs sur la mine et n'était donc pas intéressé. Il se rendit ensuite chez Peter Howe, rédacteur photo du* New York Times Magazine.

« Je me souviens que Peter a dit : "J'en fais six ou huit pages ou je quitte ce job" », raconte Dannin.

Plus de trois décennies plus tard, Howe se souvient aussi de sa réaction. « J'étais époustouflé, c'était un travail superbe, une épopée biblique, dit-il du portfolio Serra Pelada. *Je l'ai montré aux rédacteurs en chef, et là, quelque chose d'extraordinaire s'est produit. Il y a eu un silence total, le type de silence qui* travaille pour vous, pas contre vous. De toute ma carrière au New York Times, je n'ai jamais vu de rédacteurs en chef réagir ainsi à un ensemble de photographies. »

À sa grande frustration, Howe n'obtint pas la couverture du magazine du 7 juillet 1987, mais décrocha huit pages intérieures et, comme à Londres, elles eurent un impact immédiat. « La plupart des retours venaient du milieu de la photographie, explique-t-il, parce que ce n'était pas courant qu'un photographe obtienne autant de place dans le magazine pour des photos en noir et blanc. C'est comme ça que Sebastião a accédé à un niveau de prestige aux États-Unis qu'il n'avait sans doute jamais connu auparavant dans ce pays. »

Pour Dannin, chargé de commercialiser les photos de Salgado, l'impact fut plus fort encore. « Ça a révolutionné la profession, dit-il de cette saga de Serra Pelada, non seulement sur le plan de la publicité, mais aussi pour les livres. Même l'armée américaine s'est mise à recruter par des annonces publicitaires en noir et blanc. Il y a eu beaucoup d'imitations. »

En plus du magazine Life, Dannin pouvait désormais vendre ces photos à des journaux du monde entier, à la condition qu'ils s'engagent à publier des « chapitres » de La Main de l'homme au fur et à mesure de leur arrivage – ou bien à payer un dédommagement s'ils renonçaient à les publier. Beaucoup réagirent positivement, comme Paris Match et Libération en France, Stern en Allemagne, Il Venerdì di Repubblica en Italie, El País en Espagne, et bien d'autres.

La même année, en 1987, Salgado publia son dernier reportage en couleurs, sur le 70e anniversaire de la révolution russe. Depuis, parallèlement à La Main de l'homme et aux deux projets qui suivirent, Exodus et Genesis, toutes ses photographies ont paru en noir et blanc, que ce soit dans les magazines, les livres ou dans le cadre d'expositions. Allant de reportages d'illustration à des sujets autonomes, elles sont devenues des pièces de collection recherchées.

Pourquoi, alors, la photographie en noir et blanc est-elle si attrayante ?

« Après Serra Pelada, les gens se sont éveillés à la valeur de la photographie en noir et blanc, explique Neil Burgess. Pas seulement en contrepoint de la couleur, mais comme un moyen d'expression différent, une forme de photographie différente. Un photographe peut contrôler et développer sa narration plus clairement et plus directement en noir et blanc, alors que la couleur risque de distraire. C'est une esthétique différente. »

Pour Peter Howe, le noir et blanc ajoutait à la force du travail sur Serra Pelada. « Sa puissance tenait en partie au fait que cet environnement tout entier semblait complètement étranger aux Américains à la fin du XXe siècle. Il avait un aspect presque médiéval, comme s'il était surprenant de voir des *gens comme ces mineurs vivre dans de telles conditions. Le noir et blanc leur conférait également une certaine qualité, renforçant l'idée qu'il s'agissait autant d'art que de journalisme. »*

Le plus frappant pour Robert Dannin, c'était la « qualité cinématographique » des images de Serra Pelada. *« Elles font penser aux films en noir et blanc, comme ceux d'Eisenstein ou de Kurosawa, précise-t-il. Je ne dis pas que Sebastião est un réalisateur de film, mais il y a là une chorégraphie, une orchestration, dans sa manière de photographier. De même, dans* La Main de l'homme, *il semble regarder l'ère industrielle dans un rétroviseur et en noir et blanc, ce qui correspond à la façon dont nous l'avons vécue. »*

Quant à Salgado, il n'a jamais douté de sa préférence pour le noir et blanc. Dès l'enfance, quand la pâleur de sa peau le forçait à se tenir à l'écart du féroce soleil brésilien, il était fasciné par les jeux d'ombre et de lumière. Deux décennies plus tard, lorsqu'il commença à mettre au point sa technique de photographe, il comprit comment ces contrastes permettaient d'augmenter la réalité.

En revanche, il se sentait souvent distrait par la couleur. « Lorsque je regardais à travers l'objectif, je savais que les couleurs deviendraient plus importantes que les gens que je photographiais. Le noir et blanc apporte une abstraction parce qu'à travers ses tonalités et ses nuances on peut voir la dignité et la force des gens. »

C'est ce qu'illustre Serra Pelada. *Alors que les photographies en couleurs des mineurs sont dominées par la poussière ocre qui les couvre de la tête aux pieds, le noir et blanc fait ressortir chaque détail : leurs visages épuisés, leurs membres musculeux, leurs vêtements sales et mouillés. De la même façon, dans les photos panoramiques du gouffre, le noir et blanc offre des repères de lumière qui révèlent chacun des ouvriers grimpant aux échelles, chargés de sacs bourrés de terre.*

Aujourd'hui, alors que la photographie est de plus en plus absorbée par le monde de l'art et sujette à des manipulations numériques, l'aspect vintage *des images en noir et blanc contribue sans aucun doute à leur attrait. Et pourtant, pas moins que celles des maîtres du passé, les photographies en noir et blanc de Salgado communiquent un sentiment d'immédiateté qui nous les rend totalement contemporaines. Nous savons bien que la mine de Serra Pelada est maintenant fermée, et pourtant toute l'intensité dramatique de cette ruée vers l'or nous saute au visage. Comme si, en noir et blanc, le temps pouvait s'arrêter.*

Alan Riding
2019

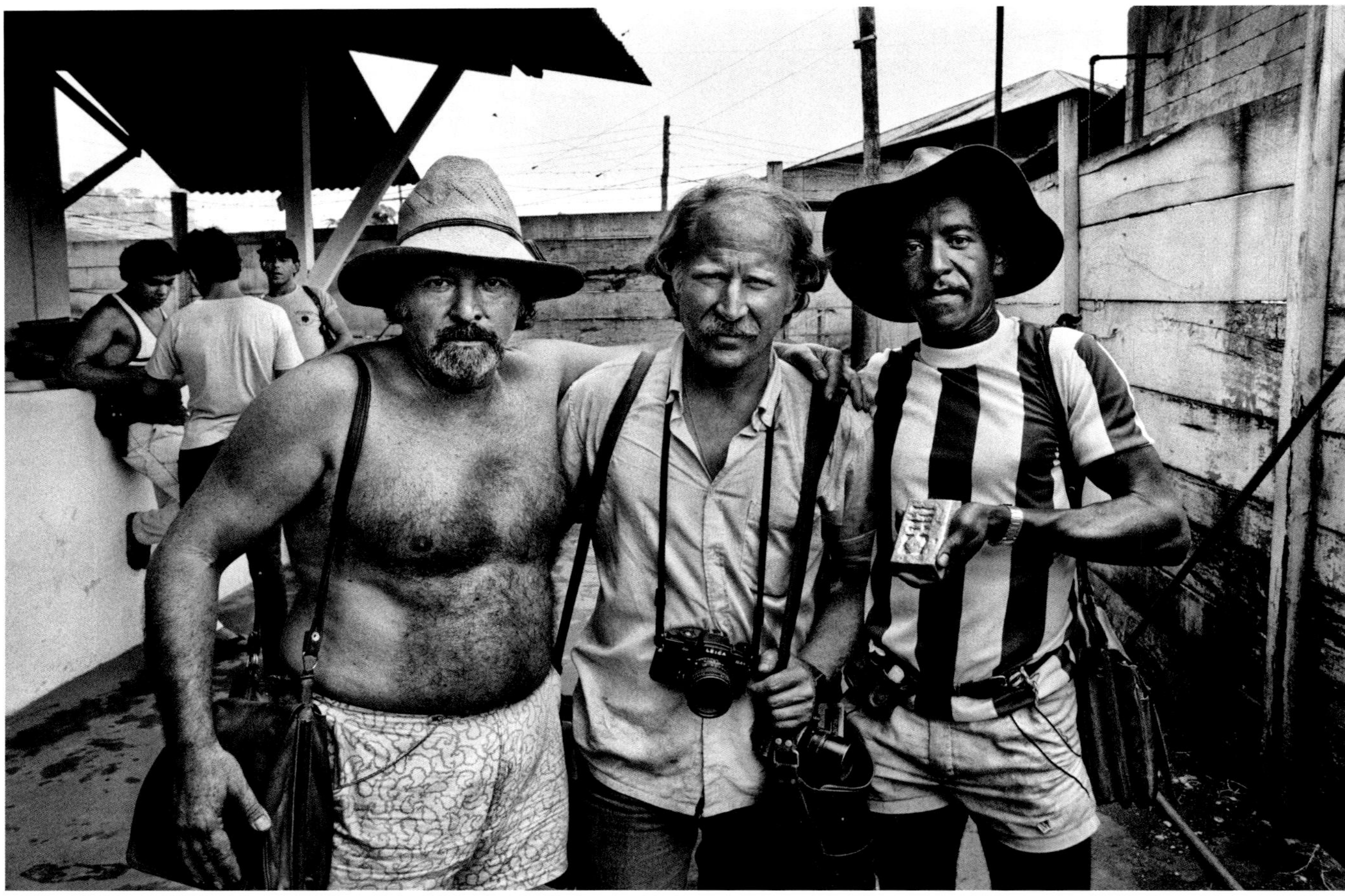

© Sebastião Salgado

Sebastião Salgado with a plot owner, holding a gold bar, and a *capitalista*, 1986.
Sebastião Salgado, der Besitzer einer Parzelle mit einem Goldbarren und ein *capitalista*, 1986.
Sebastião Salgado avec un propriétaire d'une parcelle, tenant un lingot d'or, et un capitalista, *1986.*

EACH AND EVERY TASCHEN BOOK PLANTS A SEED!
TASCHEN is a carbon neutral publisher. Each year, we offset our annual carbon emissions with carbon credits at the Instituto Terra, a reforestation program in Minas Gerais, Brazil, founded by Lélia and Sebastião Salgado. To find out more about this ecological partnership, please check: www.taschen.com/zerocarbon
Inspiration: unlimited. Carbon footprint: zero.

To stay informed about TASCHEN and our upcoming titles, please subscribe to our free magazine at www.taschen.com/magazine, follow us on Twitter, Instagram, and Facebook, or e-mail your questions to contact@taschen.com.

© 2019 TASCHEN GmbH
Hohenzollernring 53, 50672 Cologne, Germany
www.taschen.com

© 2019 SEBASTIÃO SALGADO
Photographs and Introduction Fotografien und Einleitung *Photographies et introduction*

© 2019 ALAN RIDING
Text Text *Texte*

Conception and design Konzeption und Design *Conception et réalisation*
LÉLIA WANICK SALGADO

Graphic assistant Grafikassistent *Assistant graphiste*
GÉRARD LAMARCHE

Sebastião Salgado's staff Sebastião Salgados Studioteam *Équipe de Sebastião Salgado*

Photography editor Bildredakteurin *Éditrice photo*
FRANÇOISE PIFFARD

Production Produktion *Production*
MARCIA NAVARRO MARIANO

Digital printer Digitaldrucker *Tireur numérique*
OLIVIER JAMIN

Editorial coordination Lektorat und Koordination *Coordination éditoriale*
SIMONE PHILIPPI

German translation Deutsche Übersetzung *Traduction allemande*
MICHAEL STOEBER

French translation Französische Übersetzung *Traduction française*
JACQUES BOSSER

Production coordination Herstellung *Production*
FRANK GOERHARDT, THOMAS GRELL

Printed in Italy
ISBN 978-3-8365-7508-9